Praise for *A Winter Grave*

'*A Winter Grave* is timely and chilling, painting a disturbing picture of the future . . . it's a meticulously researched thriller with gravitas that grips from the first page . . . May's first novel in two years is among the best he's written'   *Sunday Express*

'May has created a chillingly believable near future . . . an atmospheric locked room mystery'   *Observer*

'A gripping thriller set in a near future ravaged by the climate crisis'   *Scots Magazine*

'*A Winter Grave* is a superb thriller loaded with timely warnings'   *Yorkshire Post*

Praise for Peter May

'One of the glories of the modern crime novel'   *Independent*

'Peter May is a writer I'd follow to the ends of the earth'
New York Times

'A wonderfully complex book'   Peter James

Peter May was born and raised in Scotland. He was an award-winning journalist at the age of twenty-one and a published novelist at twenty-six. When his first book was adapted as a major drama series for the BBC, he quit journalism and, during the high-octane fifteen years that followed, became one of Scotland's most successful television dramatists. He created three prime-time drama series, presided over two of the highest-rated serials in his homeland as script editor and producer, and worked on more than 1,000 episodes of ratings-topping drama before deciding to leave television and return to his first love, writing novels.

He has won several literary awards in France, received the USA's Barry Award for *The Blackhouse* – the first in his internationally bestselling Lewis Trilogy; and in 2014 *Entry Island* was awarded the ITV Specsavers Crime Thriller Book Club Best Read of the Year, as well as the Deanston Scottish Crime Book of the Year. May now lives in South-West France with his wife, writer Janice Hally.

PETER MAY

A WINTER GRAVE

riverrun

First published in Great Britain in 2023
This paperback edition published in 2023 by

riverrun

an imprint of
Quercus Editions Ltd
Carmelite House
50 Victoria Embankment
London EC4Y 0DZ

An Hachette UK company

A CIP catalogue record for this book is available
from the British Library.

Paperback ISBN 978 1 52942 852 0
Ebook ISBN 978 1 52942 850 6

10 9 8 7 6 5 4 3

Typeset by CC Book Production
Printed and bound in Great Britain by Clays Ltd, Elcograf S.p.A.

Papers used by riverrun are from well-managed forests and other responsible sources.

In memory of Stephen Penn,
my best and oldest friend
1951–2022
RIP

In 1990, as NASA's Voyager 1 spacecraft was about to leave the solar system, Carl Sagan – a member of the mission's imaging team – asked that the camera be turned around to take one last look back at Earth. The image it captured of our world, as a speck less than 0.12 pixels in size, became known as 'the pale blue dot'.

Later, when considering that speck of dust in his 1994 book *Pale Blue Dot: A Vision of the Human Future in Space*, he wrote: 'There is perhaps no better demonstration of the folly of human conceits than this distant image of our tiny world. To me, it underscores our responsibility to deal more kindly with one another, and to preserve and cherish the pale blue dot, the only home we've ever known.'

PROLOGUE

NOVEMBER, 2051

Little will heighten your sense of mortality more than a confrontation with death. But right now such an encounter is the furthest thing from Addie's mind, and so she is unprepared for what is to come.

She is conflicted. Such a day as this should lift the spirits. She is almost at the summit. The wind is cold, but the sky is a crystal-clear blue, and the winter sun lays its gold across the land below. Not all of the land. Only where it rises above the shadow cast by the peaks that surround it. The loch, at its eastern end, rarely sees the sun in this mid-November. Further west, it emerges finally into sunshine, glinting a deep cut-glass blue and spangling in coruscating flashes of light. A gossamer mist hovers above its surface, almost spectral in the angled mid-morning sunshine. Recent snowfall catches the wind and is blown like dust along the ridge serpentining to the north.

But she is blind to it all. Distracted by a destiny she appears unable to change. Such things, she thinks, must be preordained. Unhappiness a natural state, broken only by rare moments of unanticipated pleasure.

The wind seems to inflate her down-filled North Face parka as well

1

as her lungs. Her daypack, with its carefully stowed flask of milky coffee and cheese sandwiches, rests lightly on her shoulders, catching the breeze a little as she turns towards the north. The peaks of the Mamores rise and fall all around her, almost every one of them a Munro, and in the distance, sunlight catches the summit of the towering Ben Nevis, the highest mountain in Scotland, the loftiest prominence in the British Isles – a little of its measured height lost now with the rise in sea levels below.

She stops here for a moment and looks back. And down. She can no longer see the tiny arcs of housing that huddle around the head of the loch where she lives. Kin is the Gaelic for head. Hence the name of the village: Kinlochleven. The settlement at the head of Loch Leven.

Somewhere away to her left lies the shimmering Blackwater Reservoir, the sweep of its dam, and the six huge black pipes laid side by side that zigzag their way down the valley to the hydro plant above the village. The occasional leak sends water under pressure fizzing into the air to make tiny rainbows where it catches the sunlight.

Finally, she focuses on the purpose of her climb. An ascent she makes once a week during the fiercest weather months of the winter to check on the condition of the flimsy little weather station she installed here – she stops to think – six years ago now. Just before she got pregnant. Fifty kilograms of metal framework and components, carried on her back in three separate trips during the more clement summer months. A tripod bolted to the rock, a central pole with sensors attached. Air temperature and relative humidity. Wind speed and direction. Ultraviolet, visible and infrared radiation. Solar panels, radio antenna, a satellite communication device. A metal box that is anchored at the summit to sandstone recrystallised into white quartzite. It contains the data logger, barometric pressure sensor, radios and battery. How it all survives here, in this most inhospitable of environments, is always a source of amazement to Addie.

It takes her less than fifteen minutes to clear the sensors of snow and ice, and to check that everything is in working order. Fifteen minutes during which she does not have to think of anything else. Fifteen minutes of escape from her depression. Fifteen minutes to forget.

When she finishes, she squats on the metal box and delves into her pack for the sandwiches thrown together in haste, and the hot, sweet coffee that will wash them down. And she cannot stop her thoughts returning to those things that have troubled her these last months. She closes her eyes, as if that might shut them out, but she carries her depression with her like the daypack on her back. If only she could shrug it from her shoulders in the same way when she returns home.

Eventually, she gets stiffly to her feet and turns towards the north-facing corrie that drops away from the curve of the summit. Coire an dà loch. The Corrie of the Two Lochans. She can see sunlight glinting on the two tiny lochs at the foot of the drop which give the corrie its name, and starts her way carefully down the west ridge. There is a mere skin of snow here, where the wind has blown it off into the corrie itself, rocks and vegetation breaking its surface like some kind of atopic dermatitis.

Before the Big Change, long-lying snow patches had become increasingly rare among the higher Scottish mountains. Thirty years ago they had all but vanished. Now they linger in the north- and east-facing corries in increasing size and number all through the summer months. Melting and freezing, melting and freezing, until they become hard like ice and impervious to the diminished estival temperatures. She had watched this patch in the Coire an dà Loch both shrink and grow across the seasons, increasing in size every year. The next snowstorm will bury it, and it will likely not be visible again until late spring.

But today there is something different about it. A yawning gap at the top end. Like the entrance to a hollow beneath it, disappearing into darkness. Maybe it had been there during her last visit, and she had simply not seen it. Obscured by snow, perhaps, which was then blown away by high winds. At any rate, she is intrigued. She has heard of snow tunnels. Periods of milder weather, as they have just experienced, sending meltwater down the corries to tunnel its way beneath the ice of long-lying snow patches.

She forgets those things that have been troubling her, and slithers down the ridge and into the corrie. The snowfall that fills this narrow valley is peppered by the rocks that break its surface from the scree below, and she has to make her way carefully across it to where the snow patch it hosts lies deep in its frozen heart. Twenty metres long, seven or eight wide. Maybe two-and-a-half deep. She arrives at the lower end of it, swinging herself round to find herself gazing up into the first snow tunnel she has ever seen. It takes her breath away. A perfect cathedral arch formed in large, geometric dimples of nascent ice stalactites above the rock and the blackened vegetation beneath it. Light from the top end of the tunnel floods down like the water before it, turning the ice blue. Big enough for her to crawl into.

She quickly removes her pack and delves into one of its pockets to retrieve her camera, then drops to her knees and climbs carefully inside. She stops several times to take photographs. Then a selfie, with the tunnel receding behind her. But she wants to capture the colour and structure of the arch, and turns on to her back so that she can shoot up and back towards the light.

The man is almost directly overhead, encased in the ice. Fully dressed, in what occurs incongruously to Addie as wholly inadequate climbing gear. He is lying face down, arms at his side, eyes and mouth wide open, staring at her for all the world as though he were still alive. But there is neither breath in his lungs, nor sight in his eyes. And Addie's scream can be heard echoing all around the Coire an dà Loch below.

CHAPTER ONE

FIVE DAYS EARLIER

The Glasgow High Court of Justiciary was an impressive building, all the more so for being stone-cleaned in the latter part of the twentieth century. A-listed as a structure of historic importance. Very few A-listers, however, had passed through its porticoed entrance. Just a long list of mostly men, in unaccustomed suits, who had gone on to wear a very different kind of attire after sentencing by the Lord Justice General, or the Lord Justice Clerk, or, more likely, one of the thirty-five Lords Commissioners of Justiciary.

Detective Inspector Cameron Brodie had given evidence in various of its courtrooms many times over the years. He was well used to the odour of the justice being dispensed by men and women in wigs and black gowns from lofty oak benches beneath artificial skylights. Justice, it seemed to him, smelled of cleaning fluid and urine and stale alcohol, with the occasional whiff of aftershave.

It was cold outside in the Saltmarket, rain leaking, as it

did most days, from a leaden sky. But the heat of legal argument in this courtroom, where a certain Jack Stalker, alias the Beanstalk, stood accused of first-degree murder, had warmed the air to a high level of humidity among all the rainwater trailed in on coats and umbrellas. Stalker sat in the dock, flanked by police officers, a grey man in his thirties with a deeply pockmarked face and a livid scar transecting his left eyebrow. Thinning hair was scraped back and plastered across the shallow slope of his skull with some evil-smelling oil that Brodie imagined he could detect from the witness stand, even above the odour of institutional justice.

Stalker's lawyer, the elderly Archibald Quayle, was well known for his defence of over five hundred murder cases, more even than the twentieth century's legendary Joe Beltrami. And despite the sweat that gathered comically in the folds of his neck and chin, he was known by Brodie to be a formidable opponent.

Quayle had wandered away from the big square table beneath the bench where the lawyers and their clerks sat, and now insinuated himself between the jury and the witness stand. He had the condescending air of a man supremely confident in his ability to achieve an acquittal, carrying about him a sense of absolute incredulity that this case had ever come to court.

To Brodie, there was no question of Stalker's guilt. He had been caught on a high-definition CCTV security camera kicking his victim to death on top of the levee on the north bank of the Clyde near the SEC conference centre.

Quayle turned dark, penetrating eyes in Brodie's direction. 'What witnesses did you interview in relation to the alleged assault, Detective Inspector?'

'None, sir.'

Quayle raised both eyebrows in mock surprise. 'And why was that?'

'We were unable to find any. The incident took place in the small hours of the morning. Apparently there was no one else in the vicinity.'

The lawyer for the defence pretended to consult his notes. 'And what forensic evidence did you acquire that led you to suspect my client of committing this heinous crime?'

'None, sir.'

The eyebrows shot up again. 'But your scenes of crime people must have gathered forensic traces from the victim and the crime scene.'

'They did.'

'Which matched nothing that you found on the accused.' A statement, not a question.

'It took us nearly two days to find Stalker. He had ample time to dispose of anything that might have linked him to the murder.'

'And how did you find him?'

'We asked around. He was known to us, sir.'

Quayle frowned. 'Known to you? How?'

Brodie took a moment before responding. He wasn't about to fall into Quayle's trap. He said evenly, 'I'm afraid that because of the Rehabilitation of Offenders Act 1974, I am unable to say how.' Which brought smiles around the lawyers' table, and a glare from the judge.

Quayle was unruffled. 'Asked around, you say. Asked who?'

'Known associates.'

'Friends, you mean?'

'Yes.'

'The victim, too, was a friend, wasn't he?'

'I believe they once shared the same accommodation.'

'Flatmates?' Quayle asked disingenuously.

Brodie paused once more. 'You might say that; I couldn't possibly comment.'

Quayle ignored the detective's flippancy and strode confidently towards his chair. 'So the only evidence you have against the accused is the CCTV footage that the advocate depute has presented to the court?'

'It's pretty damning, I think?'

'When I want your opinion, Detective Inspector, I'll ask for it.' He turned away dismissively, towards the judge. 'I wonder, my Lord, if I might ask for the court's indulgence in replaying Production Five A one more time?'

The judge glanced towards the advocate depute, who shrugged. After all, it could only reinforce the case against the accused. 'I have no objection, my Lord,' the prosecutor said.

Large screens mounted on all four walls flickered into life, and the murder of the unfortunate Archie Lafferty replayed for the umpteenth time in all its graphic detail. An argument of some kind was in progress. In full view, just across the river, of police headquarters at Pacific Quay, whose lights reflected in the dark waters of the Clyde flowing swiftly by. The levee on the north bank was deserted, except for the two antagonists. Stalker bellowed in Lafferty's face. You could almost see the spittle gathering on his lips. Then he pushed the other man in the chest with both hands and Lafferty staggered backwards, gesticulating wildly, as if pleading innocence to some savage accusation. Another push and he lost his footing, falling backwards and striking his head on the cobbles. Enough, the pathologist later confirmed, to fracture his skull, though not apparently to induce unconsciousness. Lafferty was more than aware of the kicks that rained in on him from the vicious feet of his attacker, curling up foetally to protect his head and chest. But Stalker was relentless, and when his right foot finally breached the other man's defences and caught Lafferty full in the face, you could see the spray of blood that it threw off.

The kicking continued for an inordinate and excruciating period of time, long after Lafferty had stopped trying to fend off his attacker and lay spent on the cobbles, soaking up the repeated blows and leaking blood on to stone. Stalker appeared to be enjoying himself, putting all his energy into each repeated blow, until finally he stood breathing hard and looking down on his victim with clear contempt. Lafferty was almost certainly dead by now. Stalker turned on his heel and walked briskly out of shot. The screens flickered and the video came to an end.

No matter how many times he had watched it, Brodie still felt a shiver of disquiet. A silence hung momentarily in the court, before Quayle said casually, 'That will be all, Detective Inspector.'

Brodie could barely believe it. Quayle was concluding his cross-examination with a replay of the murder, reinforcing his client's guilt in the minds of every man and woman in the courtroom. Brodie got to his feet, stepped down from the stand and walked briskly to the door.

Tiny was waiting for him outside in the hall. DI Tony Thomson was a man so thin that he didn't wear clothes, they hung on him. He measured a cool two metres, hence the nickname, and even with his voice lowered, it echoed sonorously around the tiles and painted plaster of this ancient chamber. 'That didn't take long, pal. Come on, there's a pie and a pint with our name on it at the Sarry Heid.' He turned towards the door leading to the street. But when Brodie made no move to follow, he stopped and looked back. 'What's up with you, man?'

Brodie shook his head. 'Something's not right, Tiny.'

'How?'

'Quayle had me on the stand for less than five minutes, and most of that time he spent rerunning the CCTV footage.'

Tiny frowned. 'What? He voluntarily showed the jury his client kicking shit out of that poor bastard *again*?'

Brodie nodded. 'I'm going back in.'

A few heads turned as the door creaked open and Brodie,

followed by Tiny, tiptoed into the courtroom to find them-selves places in the crowded public gallery. The advocate depute half turned and offered Brodie a quizzical frown. Brodie just shrugged.

Quayle was on his feet again. 'My Lord, I have only the one witness. I call Mr Raphael Johnson.'

The court officer returned with the witness in short order and beckoned him towards the stand. Raphael Johnson could have been no more than twenty-seven or twenty-eight years old, with a pimply, adolescent complexion and a mane of thick dark hair that tumbled over narrow shoulders. His T-shirt, beneath a hooded leather bomber, was emblazoned with the faded red logo of some unidentifiable creature breathing fire. His jeans were frayed at the knees and concertinaed over the baseball boots that were once again in fashion. Brodie clocked the nicotine-stained fingers and thumb, his bloodshot eyes and reddened nostrils betraying a likely acquaintance with a certain white powdered substance. Though perhaps Brodie was doing him an injustice. Maybe he simply had a cold, or was recovering from the latest mutation of Covid. It was hard to tell the two apart these days.

He affirmed, rather than take the oath. When asked to tell the court who he was, he called himself Raff, and described his occupation as a computer programmer with special working expertise in audiovisual manipulation.

'Who is your employer?' Quayle asked him.

'I'm self-employed, mate.'

'And your qualifications?'

'First-class honours degree in computer science from Strathclyde University.'

'Tell me about the process of video manipulation known as "deepfake".'

Raff made a snorting sound. 'No one calls it that any more, mate. Neural masking. That's what it's known as these days.'

'Tell us about it.'

The advocate depute was on his feet. 'Objection, my Lord. Relevance?'

Quayle raised a finger. 'Coming to it.'

The judge nodded. 'Be quick then, Mr Quayle.'

Quayle nodded and returned to the witness. 'Mr Johnson?'

'The technology's about thirty-five years old. Originated somewhere in the early twenty-tens, with the development of software called GAN.'

'Which is what?'

'Well, it stands for generative adversarial network, in which two neural networks use AI to out-predict one another.'

It was clear that no one in the courtroom had the least idea what he was talking about. In an attempt to be helpful, the judge leaned forward and said, 'I take it we're speaking of artificial intelligence?'

'Yes, Your Honour. It's kind of complicated to explain, but we're talking about video here, and what GANs did was produce fake videos that you really couldn't tell were fake. The two neural networks do different things. One of them is a generator; the other we call a discriminator.'

'And in layman's terms?' Quayle was hoping for more clarity.

'Well, in the early days, GAN was used to superimpose celebrity faces on to the participants in porn videos. Give the generator a few videos, or even some still samples of the celebrity face, and it would seamlessly superimpose it on to the target porn actor. You, or I, maybe couldn't tell that it had been done. But the discriminator would scan the video and find lots of faults with it. The generator would learn from that, redo the original and let the discriminator scan it again. That process would go on many times until, finally, it was virtually impossible to tell that the video wasn't genuine.'

Quayle said, 'And is it still used for that purpose?'

'Nah.' Raff shook his thick mane. 'Nobody does that any more. The software has advanced a lot since then. It has much more sophisticated applications now.'

'Such as?'

'Well, you've probably read they've started making movies with actors who've been dead for years, even decades. Big stars of the past. They employ unknown actors to make the film, then superimpose the faces of the dead stars on to them. Bingo! You've got Cary Grant playing the latest incarnation of Batman. Or Marilyn Monroe playing herself in a brand-new biopic. They can do the same thing with the voices, too. So . . .' He shrugged. 'CGI went out of business.'

Again the judge leaned forward. 'CGI?'

'Computer-generated imagery. It's how they used to turn a dozen people into a thousand in the movies, or make a scene shot in a studio seem like they were in the Bahamas. Pretty crude stuff by today's standards.'

Quayle cleared his throat and steered Raff gently back to the subject in hand. 'This neural masking,' he said. 'Just how convincing is it?'

An expression of amusement escaped Raff's lips in a tiny explosion of air. 'Mate, you can't tell it's not genuine. Unless you have the next-generation AI software – which likely won't even exist yet – there's no way to tell that it's not the real McCoy.'

Quayle nodded sagely, as if he understood every nuance of the technology being described. 'Are you able to show us an example?'

'Well, as you know, I prepared a short video by way of demonstration.'

The advocate depute was on his feet again. 'My Lord . . .'

But the judge was one step ahead of him. 'Mr Quayle, you are stretching the court's patience. This had better be good.' There was, however, no doubt in anyone's mind that his lordship was as intrigued as everyone else to see Raff's video.

'Thank you, my Lord.' Quayle nodded towards his clerk and the video screens around the courtroom flickered once more, before the video of the assault on the levee began replaying.

The judge frowned. 'That's the wrong video, Mr Quayle.'

Quayle's smile was almost imperceptible. 'No, my Lord, it's not.'

Eyes drawn by this exchange returned to the screens as Jack Stalker turned to confront his victim, and his face was caught in full street-light glare for the first time. Except that it wasn't Stalker. There was an involuntary collective gasp in the courtroom as DI Cameron Brodie's superimposed face snarled and pushed Archie Lafferty to the ground before kicking him repeatedly about the face and head. So convincing was it, that there was not a single person in the courtroom who would not have sworn that it was Brodie.

Those same eyes tore themselves away now from the video to glance at Brodie himself, sitting in the public gallery, before returning to the screens, anxious not to miss the moment. Brodie's face burned with shock and embarrassment. And anger.

CHAPTER TWO

SEVEN DAYS LATER

The rain was mixed with hail, turning to ice as it hit frozen ground and making conditions treacherous underfoot. Such little light penetrated the thick, sulphurous cloud that smothered the city, it would have been easy to mistake mid-morning for first light.

Overhead electric lights burned all the way along the corridor, making it seem even darker outside, and turning hard, cream-painted surfaces into reflective veneers that almost hurt the eyes. Brodie glanced from the windows as he strode the length of the hall. The river was swollen again and seemed sluggish as the surge from the estuary slowed its seaward passage.

The DCI's door stood ajar. Brodie could hear the distant chatter of computer keyboards and a murmur of voices from further along. They invoked a sense of hush that he was reluctant to break and he knocked softly on the door.

The voice from beyond it demonstrated no such sensitivity. 'Enter!' It was like the crack of a rifle.

17

Brodie stepped in, and Detective Chief Inspector Angus Maclaren glanced up from paperwork that lay like a snow-drift across his desk. He was in shirtsleeves, his tie loose at the neck, normally well-kempt hair falling in a loop across his forehead. He swept it back with a careless hand. 'You like a bit of hillwalking, I'm told, Brodie. Bit of climbing. That right?' There was a hint of condescension in his tone, incredulity that anyone might be drawn to indulge in such an activity. Not least one of his officers.

Born four years before the turn of the millennium, Brodie had worked his way up through the force the hard way. Graduating from Tulliallan, and spending more than ten years in uniform before sitting further exams and embarking on his investigator pathway, gaining entrance finally to the criminal investigation department as a detective constable. Two promotions later, he found himself serving under a senior officer twenty-five years his junior, who had fast-tracked his way directly to detective status as a university graduate with a degree in criminology and law from the University of Stirling. A senior officer who had little time for Brodie's *old school* approach. And even less, apparently, for his passion for hillwalking.

'Yes, sir.'

It was his widowed father, an unemployed welder made redundant from one of the last shipyards on the Clyde, who had taken him hillwalking for the first time in the West Highlands. Brodie had only been fourteen when they took the train from Queen Street up to Arrochar to climb The Cobbler, ill-dressed and ill-equipped. The right gear cost money, and his father had precious little of the stuff. But that first taste of the wild outdoors gave Brodie the bug, and as he grew more experienced, and began to earn, he started taking safety more seriously, spending all his spare time haunting sports equipment shops in the city. He was devastated when his father was struck down by a stroke. Semi-paralysed, he died a year later when Brodie was just twenty-one. And Brodie's weekend trips to the hills and mountains of the Highlands became something of an obsession, an escape from a solitary life. And in recent years, an escape from life itself.

Maclaren pushed himself back in his chair and regarded the older man speculatively. 'Remember those stories in the papers about three months ago? *Scottish Herald* reporter going missing in the West Highlands?'

Brodie didn't. 'No, sir.'

Maclaren tutted his annoyance and pushed an open folder of newspaper cuttings towards him. The *Herald* itself, the *Scotsman*, the *Record*. Most of the other national papers had gone to the wall. Apart from these, and a handful of surviving local newspapers, most people got their news from TV, internet and social media. 'A modern police officer needs to keep himself abreast of current affairs, Brodie. How can we police a society in ignorance of them?'

Brodie supposed that the question was rhetorical and maintained a silence that drew a look from Maclaren, as if he suspected dumb insolence.

'Charles Younger,' he said. 'The paper's investigative reporter. Specialised in political scandals. Last August he went hillwalking in the Loch Leven area, even though by all accounts he'd never been hillwalking in his life. Went out one day, never came back. No trace of him ever found. Until now.' He paused, as if waiting for Brodie to ask. When he didn't, the DCI sighed impatiently and added, 'Younger's body was discovered frozen in a snow patch in a north-facing corrie of Binnein Mòr, above the village of—'

Brodie interrupted for the first time. 'I know where Binnein Mòr is. I've climbed most of the mountains in the Kinlochleven area.'

'Aye, so I heard. All of the Munros in the Mamores, I believe.'

Brodie offered a single nod in affirmation.

'I want you to go up there and check it out.'

'Why are Inverness not dealing with it?'

'Because the two officers they sent to investigate were killed when their drone came down in an ice storm. Edinburgh have asked us to send someone instead. And I'm asking you.'

'Then you'll have to ask someone else, sir.'

Maclaren canted his head and Brodie saw coals of anger stoking themselves in his eyes. 'And why the fuck would I do that?'

'I have a doctor's appointment today, sir. To get the result of hospital tests. I'm likely to require treatment.'

Maclaren glared at him for a moment, before banging the cuttings folder closed and drawing it back towards himself.

'Why didn't I know about this?' No concern or query about the state of his health.

Brodie said, 'You will, sir. When I've got something to tell you.' He glanced at his watch. 'If that's all, sir, I have to go. There's a tech briefing at ten-thirty, and I wouldn't like to keep the DCS waiting.'

Tiny joined his long-time partner as he stepped into the lift and pressed the button for the fifth floor. 'So what did you tell him?'

'To go fuck himself.'

Tiny pulled a face. 'Aye, right. What did you really tell him?'

'That I didn't want to do it.'

'I thought you'd have jumped at the chance, pal. Right up your street, that. Climbing mountains and shit.'

Brodie shrugged. He wasn't about to go into medical details, even with his oldest friend.

'Anyway, I thought your daughter lived in Kinlochleven these days.'

Brodie nodded.

'So . . .'

'So, maybe that's why I don't want to go.'

The lift doors slid open and Brodie stepped briskly out into the hall. Tiny fell into step beside him as they walked along to the briefing room at the end of the corridor, and held his tongue. He knew better than to push Brodie on the touchy subject of his daughter.

The briefing room was packed with both uniformed and plain-clothes officers. This was to be the much-anticipated introduction of the new comms kit, some kind of ultralight mobile video phone that packed more processing power than most desktop computers. Everyone was anxious to get a sight of it. And get their hands on one.

Brodie and Tiny found seats by the window and Brodie looked out across the Clyde. The new Glasgow police HQ had been built at Pacific Quay in the early thirties and, like its neighbouring media complexes – the publicly owned Scottish Broadcasting Corporation, and the commercial Scottish Television – it stood hemmed in by the levees built in the forties to provide protection against the storm and tidal surges that had flooded large areas south of the river. Despite radical changes to local government since the country voted for independence in the late twenties, Police Scotland was still a unitary force.

After the country's reaccession to the European Union, sponsored by France, the new Scottish Government at Holyrood had restructured largely along the French model. Scotland was divided now into four regions, roughly corresponding to the country's diagonal geological fault lines – Central, South, Mid and Highland – then carved up into departments administered by government appointees. These were subdivided, with towns and defined rural areas being established as cantons, each electing its own local mayor. Both the Western and the Northern Isles had been declared separate, semi-autonomous mini-regions.

So much had changed in Brodie's lifetime that he found it hard to keep up, and harder still to summon any interest in doing so.

Rain ran down the windowpanes, distorting the outline of the Armadillo across the river. The Finnieston Crane was almost obscured by it. Distant tenements, standing atop the hill that rose above Partick, where he lived, were a depressing rust-red blur, almost subsumed, somehow, by the sky.

A hush of anticipation fell across the gathering when the DCS strode in. He was followed by a bespectacled younger man in plain clothes, with hair that grew down almost to his collar. He was carrying a large cardboard box stencilled with the logo 'iCom'. Both installed themselves behind a desk that sat below the whiteboard on the end wall, and the box was placed on top of it. The DCS removed his chequered hat and laid it on the table in front of him. He had a thick head of silvered hair, and a shiny, shaven, well-defined jawline. About the same age as himself, Brodie reckoned. But today there was something different about him.

And as if reading his thoughts, the DCS said, 'How many of you have noticed a change in my appearance today?'

Tiny called out, 'You're wearing glasses, sir.' He'd always had an eye for detail that Brodie envied.

'Correct, Thomson. And yet, not.' There was a moment's puzzled silence. 'They are not glasses.' He raised a hand to one of the legs and removed them from his face, leaving the elements that curl around the ear in place. 'Believe it or not, the legs of a pair of spectacles are called *temples*. In these iCom units, the temples detach from the ends that loop around your ear, and reattach magnetically. You can take them off like this, or you can swivel them up on to your forehead.' He placed the glasses back on his nose, then pushed them up into his hairline to demonstrate. 'If I ask my iCom to darken the lenses, then I look like I'm wearing sunglasses.' He adopted a commanding tone. 'iCom, shade my lenses.' They instantly turned dark as he dropped them over his eyes again. 'There. Now I look cool, right?'

'Poser,' somebody said in a stage whisper, which drew laughter from around the room. The DCS grinned, anxious to show that he, too, had a sense of humour.

Raising a hand to his right ear, he said, 'The piece that goes in and around your ear on each side translates sound into silent vibration that your brain then retranslates into sound. It's very sharp, very clear, and no one can hear it but you.' He ran his index finger from the back of his ear around the curved end of his jaw below it. 'You probably can't even see this, because it's flesh-coloured and will adapt to the tone of your skin, whatever that might be. But it picks up the vibration from your jaw as you speak and sends it as a voice signal across the police 15G network. So you will be in constant two-way communication with whoever you call.' He pushed the glasses back into his hairline. 'Bring the glasses into play, and they provide an augmented reality VR screen that allows you to receive video calls, surf the internet, or interpret the world around you. Facial recognition is instant. Everything functions on voice command.' He smiled. 'But here's the beauty of it: you can still see everything that's going on beyond the lenses. It's just a matter of jumping focus. You get used to it very quickly.'

'What about two-way video?' someone said.

The DCS turned to the younger man standing impatiently beside him. 'This is DI Victor Graham from IT. Our hacker in chief.' The hacker in chief seemed less than impressed by his monicker. 'He can explain it better than me.'

Graham removed his own glasses and ran a delicate finger around the outside edge of the lenses. 'There are eight tiny cameras built into the rims,' he said. 'They scan your face and reinterpret the digital information to send a faithful video rendering of your likeness to the other party.' He replaced his glasses. 'Make no mistake, the processing power of these iCom sets is enormous, powered by miniature cells built into the end pieces.' He touched the angled joints where the legs were hinged to the frames. 'You'll get about ninety-six hours of uninterrupted use without having to recharge.'

The DCS stepped in again. 'Now here's a really interesting feature . . .' he smiled, 'which should appeal to our friend, DI Brodie.'

Heads turned towards Brodie, and he felt the colour rising on his cheeks.

'Software in the iCom will allow officers to view video and scan it to determine whether or not it is genuine.'

Graham said, 'The process is lighting-fast, and the software is generations ahead of the competition. It's foolproof.'

The DCS grinned. 'So you'll all be able to tell whether the actress in your porn videos is real or not.' Which brought a ripple of laughter around the room. 'Just a pity it wasn't available last week when Brodie fucked up the case against Jack Stalker. Bastard wouldn't have walked free, then, eh?'

Brodie clenched his jaw.

'Okay, I'm going to hand you over to DI Graham here to provide a full briefing and issue you with your individual iComs. Any queries, direct them to him. Lose the fucking thing and you'll answer to me.'

He picked up his hat to set squarely on his head and marched briskly out of the room.

DI Graham waited until he was gone. 'And me,' he said. 'These things come out of my budget, and they cost a fucking fortune.'

CHAPTER THREE

An air-conditioning fan rattled and whirred behind the rusted grill in the ceiling. Rain thundered on a skylight that spilled bruised daylight into the waiting room. The sound of the large-screen television fixed high up on the far wall was only just audible above it. Discoloured plastic chairs stood lined up against three walls, facing a low square table in the centre of the room. The table groaned with grubby, dog-eared magazines, and Brodie imagined them to be contaminated by the invisible bacterial and viral infections carried by all of the sick patients who had handled them.

The walls of the room had not been painted in years, and were stained with damp and scarred by the backs of chairs. It was empty when Brodie first entered, dripping rainwater on the floor after a perilous ride through flooded streets in the open electric taxi boat he had caught at one of the temporary south-side jetties. Private boats for hire clustered around all the jetties like so many feeding fish.

He was always depressed by the rain-streaked sandstone tenements that lined the streets. They stood between the gap sites like the few remaining rotten teeth in a sad smile. Abandoned like the tower blocks and the newer social housing. Shop windows had been boarded up long ago, and were almost obscured by graffiti. The Citizens Theatre in Gorbals Street had been forced to close its doors permanently after almost a hundred years of productions on the stage of what had once been known as the Royal Princess's Theatre. All the drama these days played out on water in the streets around it.

For a while he had sat on his own in the waiting room, feeling the air thicken with humidity, before an elderly man in a flat cap and dripping grey raincoat pushed open the door and took a seat against the far wall. After the briefest nod of acknowledgement, he had begun amusing himself by stamping on the cockroaches scuttling across the tiles. The hardy German variety of the insect that infested the city had moved indoors to survive the falling temperatures which had come unexpectedly with climate change. The little bastards were hard to kill. Brodie watched, fascinated, for a while, before finding himself drawn by a familiar jingle interrupting a succession of annoying infomercials on the television. The equally irritating jingle was the one adopted by the Eco Party to herald its endless political party broadcasts ahead of the imminent election.

The incumbent Scottish Democratic Party, led by the charismatic Sally Mack, was well ahead in the polls. The SDP, unlike the EP, did not seem to feel the need to constantly badger the electorate for their votes. Which imbued them, somehow, with a reassuring sense of self-confidence, even superiority.

The Scottish Tories had long since faded into oblivion, leaving the Ecologists as the only genuine opposition. But there was a sense of desperation in their floundering campaign as election day approached.

Their latest offering was a rerun of the testimony given to a US Senate committee by the famous twentieth-century American scientist Carl Sagan in 1985. Dark hair, greying at the temples, fell carelessly around his large skull. His face was dominated by huge teardrop glasses, a reflection perhaps of his fear for the future. But his voice was almost soporifically calm, despite the tenor of his subject. Climate change. A favourite topic of the Eco Party. A concern, Brodie thought, that was thirty years and more too late. In fact, more than twice that, if Sagan was to be believed.

In his evidence on climate change, he told the senators, 'Because the effects occupy more than a human generation, there is a tendency to say that they are not our problem. Of course, then, they are nobody's problem. *Not on my tour of duty. Not on my term of office.* It's something for the next century. Let the next century worry about it.'

Brodie shook his head. They were halfway through the next century, and the fact that nobody had done nearly enough worrying about it was self-evident.

'And so,' Sagan went on, 'in this issue, as in so many other issues, we are passing on extremely grave problems to our children, when the time to solve the problems, if they can be solved at all, is now.'

Brodie could barely hear him above the rain hammering on the skylight.

'The solution to this problem requires a perspective that embraces the planet and the future, because we are all in this greenhouse together.'

Out of interest, Brodie slipped on his new glasses. He felt the magnets lock into place as the legs connected with the earpieces he and his fellow officers had been asked to wear while on duty, and he requested his iCom to scan the Sagan video for authenticity. The old man on the other side of the waiting room looked up momentarily from his cockroach squashing, and wondered who Brodie was talking to.

As his iCom performed its scan, Brodie noticed a cockroach crawling across the lower portion of Sagan's face, reaching his lips as he spoke. Brodie almost expected it to disappear into his mouth, choking off the words of warning. *Scan completed,* flashed up on his screen. *Video authenticated.* So, the Eco Party was correct in its assertion that the world had been given notice more than sixty years ago.

A grim-faced Eco spokesman appeared on-screen, urging voters to crush the Democrats at the polls, as if somehow the climate change afflicting them had been brought about solely by the SDP.

Brodie pushed the glasses up on to his forehead and sighed his frustration. He hated politics. Politicians all told lies. Lies that changed depending on what demographic they were appealing to.

The sudden opening of the door from the doctor's surgery startled him back to the grim reality of the waiting room. A middle-aged woman, head bowed, hurried past and out into the hall to negotiate the gloomy curve of the stairs down to the flooded streets below.

The doctor was a good ten years younger than Brodie and almost completely bald. He wore a tweed suit and horn-rimmed glasses, and waved Brodie into his inner sanctum. He beckoned the policeman absently towards a seat, closed the door and rounded his desk, distracted.

'Fucking cockroaches,' he said, and Brodie blinked in surprise. The doctor never once looked at him as he shuffled through the plethora of papers on his desk. 'The extermination people were supposed to come last week. We're overrun with the damn things. Little fuckers are everywhere. This is a doctor's surgery, for Christ's sake. It's supposed to be a sanitary environment.' He looked up for the first time. 'Do you have any idea what kind of diseases are carried by cockroaches?'

Brodie didn't.

'Dysentery, gastroenteritis, salmonella . . .' The doctor waved frustrated arms and slumped into his seat. 'It's a bloody disgrace.' Scanning his desk again, he pulled a folder towards him to open it up and examine the contents. He rubbed the bristles on his chin and Brodie heard the scrape of them against the soft skin of his palm. He turned gloomy eyes towards his patient. 'It's bad news, I'm afraid.'

*

Brodie emerged from the doctor's surgery, back into the waiting room, like a man in a trance. As if every nerve in his body had suddenly surrendered perception. He had no sense of putting one foot in front of the other. Breathing had become a conscious act that required concentration. The tinnitus in his ears drowned out the world.

The old man in the flat cap and raincoat pushed past him in his hurry to enter the surgery, as if someone might suddenly appear to jump his place in a non-existent queue. Brodie heard the door behind him close. He stopped, standing in the waiting room below the green-blue light of the window above. He was not conscious of thinking about anything.

Then, incongruously, he became aware of the Carl Sagan interview replaying on the TV. *We are passing on extremely grave problems to our children.* And his eyes flickered towards the screen. The remains of the cockroach were smeared across his mouth. Its brown innards and smashed wings stuck to a rolled-up magazine lying on the table.

Rain still battered the glass overhead. The air-conditioning fan still whirred and rattled behind its rusted grill. Nothing had changed. Except everything had.

Outside, he saw a group of water taxis gathered beneath a cluster of umbrellas at the entrance to the waterlogged car park of the Central Mosque. The drivers were playing cards under black protective oiled cloth, and only by shouting was he able to attract their attention. One of them reluctantly disengaged, swinging his tiller and directing his shallow-draft boat silently in the direction of the medical centre.

'Where you going, pal?'

'Suspension bridge, Carlton Place.'

The driver breathed his frustration. 'Hardly worth the fucking fare.'

'How else am I supposed to get there?'

He shook his head. 'Get in.'

Brodie clambered into the front of the boat and sat watching the buildings drift by in the rain. The mosque had been closed for several years. The underground, once known as the Clockwork Orange because of its single circle and orange trains, had been flooded in the first storm surges and never reopened. Nearby shops and apartments had been ruthlessly looted in the early days of the initial flooding. And although there was no longer anything of value left to steal, these were still dangerous streets after dark. Gangs of white youths roamed in high-powered boats looking for trouble, searching out the Asian immigrants who had poured into the country over the last decade to colonise large parts of this cold northern city, escaping disastrous flooding and crop failure on the subcontinent.

But, in truth, he saw nothing but the fading light of his own mortality. Of course, we were all going to die sometime. But it is a very human trait to lock that thought away, to face it when keeping it in the dark is no longer viable. You know that one day it will seek you out, but are never prepared for it when it does. The doctor's words still rattled about in his head.

You have severe and rapid-moving prostate cancer. Unfortunately, it has metasticised to the bone. Ribs, hips, the small intestine . . .

Strangely, except for the blood in his urine that had prompted the first visit to the doctor, he felt fine. There were the occasional bouts of fatigue, and the difficulty he had some nights in sleeping. But, then, he had never been a good sleeper. Not since that long ago night in the dark under the King George V Bridge on the Clyde.

His boat turned into Carlton Place. Falling behind in the building of levees, the Glasgow prefecture had belatedly undertaken work to raise the entire suspension bridge, which still only just cleared the water during the worst storm surges. The disgruntled driver dropped Brodie at the foot of the steps and examined the handful of coins placed in his hand. 'Make more at the fucking poker,' he said, and spun his boat away to hurry back to the game, sending a wash of black water sloshing into the flooded basements of the buildings that lined this once grand terrace. The Sheriff Court at the end of the street had been built above street level, with steps leading up to its entrance. Opened in 1986, it was almost as if the architect who designed it had taken heed of Carl Sagan's warning and placed the administration of justice beyond the reach of rising sea levels and storm surges.

Brodie walked across the bridge, the rain slashing his face. It was ice-cold, and numbed his skin as effectively as the doctor's words had numbed his senses. There was no room for fear. Just a yawning, aching emptiness.

CHAPTER FOUR

Brodie's top-floor flat looked out across the city, east and west, from this red sandstone tenement at the top end of Gardner Street.

If he looked out the back from his kitchen window, he could see down into the cricket ground at Hamilton Crescent, where the first Scotland–England international football match had taken place in 1872. A goalless draw played out in front of 4,000 spectators.

From the bay window at the front, he could look down the hill to Dumbarton Road, which flooded frequently when the sewers and drainage system backed up from the river. The smell that lingered along the famous old road was nauseating. He had no idea how people continued to live there, how the shops survived. He avoided it as far as it was possible. Fortunately the odour rarely made it this far up.

It was dark by the time he got home. He switched on a light above the kitchen sink, and saw dirty crockery soaking in scummy water. Pans with the burned-on remains of forgotten meals. Aluminium carry-out cartons were stacked untidily on the worktop next to piles of cardboard lids stained with the residue of Chinese and Indian takeaways. The stink of stale food was very nearly as odious as the smell that stalked Dumbarton Road.

He saw his face reflected in the glass of the window. A ghost of himself. Short hair bristling silver and black and receding from a high brow above blue eyes. Cheeks hollowed out in the shadows cast by the light, and by the words of death delivered by a doctor more concerned with an infestation of cockroaches.

Unable to look at himself, he turned off the light and went through to the front room. He had no appetite, either for eating or for clearing up the mess. Neither seemed to have any point any more. Light from the street lamp outside reflected on the ceiling, casting the shadows of furniture around the room. He dropped heavily on to the settee. A half-empty bottle of whisky nestled between the arm and the cushion on his right. Golden oblivion. He reached for it and unscrewed the cap, raising the neck of it to his lips and letting the amber liquid burn all the way down to his gut. Maybe if he drank enough of it, the alcohol would kill the fucking cancer. Or him. Either would do.

It took him no time at all to empty the bottle, but if he had hoped to find escape in it, he was disappointed. Oblivion evaded him. When he closed his eyes, the room spun, and when he opened them again, the world was still there. Unchanged. Dark and depressing. The only emotion he seemed able to conjure from the whisky was self-pity. He forced himself out of the settee and staggered across the room to a G-plan sideboard that his parents had inherited from his grandparents, and then by him from them. The drawers were full of detritus from his folks' house that he had never bothered to clear out, and when opened, always released a flood of memories in the timeless smell that lingered still among their belongings. He lifted out an ancient, dark green photo album embossed with a crocodile-skin pattern. Its brittle pages were held in place by red string threaded through punched holes. *Photographs*, read the washed-out gold script in the bottom right of the front cover. Between each page of photographs, as he squinted at them in the semi-dark, lay a protective sheet of patterned greaseproof paper.

The album had belonged to his paternal grandparents and dated back to a generation before that, black and white prints lovingly pasted on to each page, annotated in faded black ink. *Granny Black. Papa Brodie.* People long dead, like everyone else pressed between these pages. Lives that had come and gone, genes passing from one generation to the next in a line that ended with him. And he wondered what the hell it was all about. What any of it meant. What any of it was for.

And then he realised, of course, that it didn't all end with him. That was just self-pity, self-obsession. There was Addie. Who wouldn't even speak to him. If procreation was your raison d'être, then surely the estrangement of your child was the ultimate failure.

He dropped the album back in the drawer and went in search of another bottle. There was a little gin left in one he found in the kitchen, and he brought it back with him to the settee. It tasted vile after the whisky and he spat it out in an aerosol spray that settled slowly in the still of the room, the finest droplets catching light from the window as they fell.

'Fuck!' His voice reverberated around the room. Then, 'TV, show me my photos.'

The screen on the wall above the original Edwardian fireplace flashed and flickered before presenting a pale blue background to a succession of digital photographs. They arrived like postage stamps from one side, grew to fill the screen, then shrank to vanish from the other. An accumulation of pictures taken in happier times, increasing in quality with each new generation of iPhone, at almost the same rate as the people in them grew older. And sadder.

His heart ached as it did every time he set eyes on Mel again. So young and fresh at first. Beautiful in her plainness. A wisp of a girl with a smile that would break your heart. Big, doleful eyes hiding mischief in their darkness. Long, straight, mouse-brown hair tonged into curls for their wedding day. He was transfixed by her, unable even to look at the pictures of himself, afraid that he would see only what he had become, not what he had been. Thirty seconds of video scrolled by. Mel laughing, almost choking on a piece of wedding cake. 'Try it, Cam,' she was saying and holding out cake towards the camera. 'Try it.' So innocent, and so cursed. And long dead. Like his parents, and their parents before them.

CHAPTER FIVE

2023

In those days I couldn't see myself ever getting married. I was twenty-seven years old, and I knew a lot of guys that age who were still living with their parents, drawing on the bank of Mum and Dad. But I had a top-floor flat of my own up in Maryhill. A great view of the cemetery, and a problem with dry rot that seemed to be creeping its way through the building. But it was just a rental, so what did I care?

Tiny and I had shared it in the years after we graduated from Tulliallan. That's the police college in Fife. It's where we met. Hit it off straight away. And blazed our way through all the pubs and clubs in Dunfermline. Heading into Edinburgh on the weekends to try our luck with the capital girls.

We were lucky to finish our training together in the Glasgow East command area, posted to the same police station in London Road. We weren't earning much in those days, so a one-bedroom tenement flat was all we could afford. We took it in turns, alternating weeks, to sleep on the sofa bed in the front room. It was better than staying with your folks. Not that I had any left by then. But Tiny's people were still alive.

Everyone thought there was an endless stream of girls passing through that wee flat. But in truth it was usually just me and Tiny, a six-pack of Tennent's Lager and Sky Sports on the telly.

He was a good pal, Tiny. You could tell him anything and trust him to keep it to himself. Made me laugh. Big, long drink of water that he was. Always had a running commentary on life. You know, that observational thing that Billy Connolly had. An eye for the absurd. Always saw the funny side, even in the darkest moments.

I can confess to it now, even if I couldn't admit it to myself at the time, but I was devastated when he met Sheila. Suddenly all those nights and weekends when we'd be watching the game, or down the pub, or off clubbing in town, came to an end. It was like losing a leg. I'd never really thought about the future. I guess I didn't want to. I was happy with the life we had, the respect we got as cops (usually). Someone to share my thoughts with and have a laugh together. Fuck's sake, it was almost like we were married!

Then everything was about Sheila. *Can't go to the pub. Can't go to the game. Me and Sheila are going to the flicks tonight. Sheila's booked a table at that Chinese in Hope Street. I'd ask you to join us, but, you know . . .*

That's when I got really serious about the hillwalking. If I was going to be on my own, I'd rather be climbing a mountain somewhere than sitting on my tod fetching endless beers from the fridge and watching a game that was only half as entertaining without the banter.

I was jealous as fuck when he said they were getting married. Of course, I agreed to be his best man. How could I tell him I didn't really like his Sheila very much? I suppose there was never any chance I would. After all, she'd stolen my best mate.

He and Sheila put down a deposit on one of those four-in-a-block houses in King's Park, and after the wedding I hardly saw him outside of work. To be honest, I didn't want to socialise with the two of them, and I'm pretty sure Sheila didn't like me very much anyway. I stayed on at the flat after Tiny left, but I was spending less and less time there. Every day I wasn't working, every holiday, I was off up the Highlands bagging a Munro. In Scotland, that's any mountain over 3,000 feet. There are 282 of them, and I must have clocked up well over a hundred back in the day. In the Mamores, the Cairngorms, the Grampians . . .

Tiny was still my mate, always will be. But it wasn't the same any more. We were cops together and that was it. And I was sick to death of him telling me how great it was being married, and how I needed to find myself a woman and settle down, raise a family. Irony of it is, I was the one that ended up having the kid. Tiny and Sheila never could.

It all changed for me one October night, about a year after they were married. Tiny and I were still working out of London

Road. We were lucky. We had a BMW 530, which could fairly shift when we needed it to. Tiny usually drove, cos he had these long legs that meant he had to push the seat right back, and I couldn't be bothered readjusting it every time. I mean, I'm not short. Just under six foot. But my feet wouldn't even reach the pedals.

A call came over the radio for us to attend a domestic at a block of flats at Soutra Place in Cranhill. Overlooking that tousy wee park. Routine shit. It was pitch when we got there, lights in all three towers burning against a black sky that had been spitting rain at us all night. Seventeen storeys in those blocks. It was just our luck that the domestic was on the fifteenth and the fucking lift wasn't working.

I was used to climbing, so it didn't really bother me. But Tiny was well out of puff by the time we got there. And we could hear the raised voices all the way down the hall. It sounded like World War III. The man's voice dominating, and what sounded like a young girl pleading. A constant stream of exhortations for him to stop. And then a scream when he hit her. There were other residents standing in open doorways as we pushed along to the end door.

'Took yer fucking time,' someone told us in a voice that sounded like sandpaper.

'It's been going on for hours,' a woman said. 'He's going to kill her one of these days!'

Tiny hammered on the door, and the sound of it reverberated all the way back down the hall. There was a sudden silence inside. A moment. Then a man's voice shouting, 'What the fuck?'

Tiny glanced at the nameplate above the bell. 'Open up, Mr Jardine, it's the police.'

Another pause. 'Fuck off!'

My turn. 'Sir, we need to verify that there is not a criminal assault in progress. If you don't open up, we're going to have to call in reinforcements and break your door down.'

The door flew open and Jardine stood silhouetted against the light in the hall behind him, swaying unsteadily. The smell of alcohol off him was rank. He was a big man. Not as tall as Tiny, but built. He had a half-grown beard on a pale face that was oddly handsome in its own way. Green eyes that seemed lit from behind. Sculpted eyebrows and a shock of thick, black hair. 'There,' he said. 'I'm fucking fine. See? Nae blood.'

I peered around him, trying to see into the living room. 'And the young lady?'

'That's no lady, that's my wife. Ha, ha, ha. Only joking. She's my bidey-in and she's fine. Alright?'

'We'd like to verify that, sir,' Tiny said, and Jardine found himself looking up into Tiny's implacable face. Probably something he really wasn't used to.

Without a word he stood aside, holding the door open, and we went through into a room that looked like a bomb had gone off in it. Chairs were overturned. A burst cushion had sent feathers flying. They were still settling. There was an overturned wine bottle and a broken tumbler on a coffee table that was scorched and pitted by cigarette burns. The place smelled of alcohol and vomit and stale smoke, a fugg of it still hanging in the air. An overhead lamp threw a cold yellow light on to this sad scene of domestic bliss, casting cruel shadows on the slip of a girl who sat on the settee, hunched forward, palms pressed together between her knees.

It was the first time I ever set eyes on Mel. And I guess I knew even then there was something special about her. Can't say what it was. I mean, she was no beauty. Not in any conventional way. There wasn't a trace of make-up on a face that was swollen and bruised, blood clotting on a split lip. Her hair was greasy and limp, and hanging down like hanks of torn curtain that she was trying to draw on herself. As if somehow they could hide her shame.

I suppose it was her eyes. I'd never seen eyes that dark. I'd read descriptions in cheap novels of folk having eyes like coal, but it was the first time I'd been able to picture it. Later I understood that while her eyes really were a very deep brown, it was the dilation of her pupils that had made them so black that night. But you could see there was light in the darkness. And something that said there was intelligence there too, even if it wasn't immediately apparent.

She wore a bloodstained T-shirt and baggy blue jog pants, bare feet revealing pale pink painted toenails with chipped and broken varnish.

I figured she was eighteen, maybe nineteen, and couldn't work out what she was doing with a man a good ten years her senior. A brute of a man at that. My first instinct was to lift her to her feet and take her in my arms. My second was to beat the shit out of the man who'd done this to her. I did neither.

Tiny said, 'Big man, eh? Beating up on a lassie.'

'She fell,' Jardine said.

I couldn't bring myself to speak, but the look on my face must have said it all.

He stared back at me. 'What!'

I said, 'I think you'd better come down to the station with us, Mr Jardine.'

'And why would I do that?'

'The desk sergeant might just want to charge you with breach of the peace.'

'Whose fucking peace?'

'The peace of every neighbour who called to complain about the noise you've been making.'

Tiny said, 'And then there's the question of assault.' He reached for Jardine's wrist, but wasn't expecting the reaction he got.

Jardine pulled away abruptly, fuelled by that potent mix of alcohol and the anger that burns in all bullies. 'I'll give you fucking assault,' he shouted. And his clenched fist came swinging into the room. A punch that connected with nothing but fresh air as Tiny took a step back and Jardine lost his balance.

I moved in fast, catching his forearm and swinging him round to bang face-first into the wall. Tiny had the cuffs on him before he could move. 'Add resisting arrest to that count,' I breathed into his ear.

Tiny took him down to the car then, and I stayed with the girl to see if she needed medical attention.

'I'm alright,' she insisted. But I made her sit where she was and went through to the kitchen to boil a kettle, then wadded up some kitchen roll to clean the blood from a cut in her hairline and the split on her lip. She pulled away from the sting of it, and I looked at the bruising coming up on her face. I could see that there was old yellow bruising beneath the fresh stuff, and more on her forearms, where maybe she had raised them to protect herself.

'You should put witch hazel on that bruising,' I said. I remembered that my mum had always kept some in the house for when I had a tumble from my bike or got into a fight at school.

Unexpectedly, she laughed, and her face shone as if someone had turned on a light. 'Don't know any witches,' she said. 'And for sure not any called Hazel.'

'I'll need to introduce you to her, then. She can magic those bruises away.' She smiled and I said, 'What are you doing with him, darling?' And the light went out. Tears filled those dark eyes and she shook her head.

'None of your business.'

And maybe it wasn't. We sat for a moment before I said, 'What's your name?'

In a tiny voice she said, 'Mel.' And then smiled again through her tears. 'My mum was a big Spice Girls fan.'

Of course I'd heard of the Spice Girls, but they'd gone their separate ways before I even started primary school, so the reference was kind of lost on me at the time.

She saw my confusion and grinned. 'Two of them were called Mel. Well, Melanie, I suppose. But I was just Mel. That's what's on my birth certificate. Just plain old Mel.'

It's hard to describe why, but something in the childlike innocence of this touched me. It was a quality she had that she never lost, and that never failed to affect me. I would learn in time that she was also smart, and perceptive. But it was that intractable innocence that led to her downfall.

It was still raining when I got to the car. I don't know why, but I had kind of worked myself up into a lather going back down all those stairs. The only thing I could picture was that pale bruised face, and the innocence of her smile. And the drunken fist that Jardine had thrown at Tiny. The thought of it connecting with Mel. I knew that whatever happened to him tonight, he would take it out on her when he got home, and I wanted him to know I wasn't about to let that happen.

Tiny was sitting at the wheel with the window down. 'She alright, mate?' he said. But I just walked past and opened the rear door. Jardine wasn't expecting it, so it was easy enough to pull him out on to the forecourt. He fell to his knees before scrambling unsteadily to his feet. I heard Tiny's voice from somewhere behind me. 'What the fuck?'

I grabbed Jardine's jacket and pushed him up against the

car, thrusting my face in his. 'Lay a finger on that lassie again, Jardine, and I'll fucking have you.'

'You and whose fucking army?' he roared. And I was totally unprepared for the headbutt. A Glasgow kiss delivered properly will break your nose, but all that Jardine managed was a clash of foreheads that stunned him and infuriated me.

I piled in with knees and fists, catching him in the crotch and pummelling his ribs until his legs gave way. A final fist caught him full in the face, jerking his head to the side before he vomited on the tarmac.

Tiny was pulling me away, his voice hissing in the dark, 'Jesus Christ, man! Stop!

I turned towards him. 'He headbutted me. You saw that, didn't you?'

His face was dark with anger. 'Fuck's sake, Cammie! Get in the fucking car.' And he dragged Jardine to his feet and bundled him in the back.

London Road police station comprised a long, three-storey brick building that stood in an industrial desert in the east end of Glasgow, a spit away from Celtic Park football ground. The compound at the back of it housed umpteen overspill Portakabins that had become permanent fixtures. It was a depressing place at the best of times.

The sergeant behind the charge bar cast a dubious eye over the sorry figure we presented to him at a little after one o'clock that morning. From the driving licence in Jardine's wallet, we had gleaned that his full name was Lee Alexander Jardine, and that he was thirty-one years of age.

The blood from his nose had dried on his face, with one eye bruised and puffed up till it was almost closed. I figured there was probably a loose tooth or two, but that wasn't obvious at a glance. His wrists were still cuffed behind him, and he stood half-hunched, his jacket stained with his own vomit, the stink of alcohol hanging about him in a cloud.

The sergeant swivelled his eyes in my direction and took in the swelling on my forehead. 'Breach of the peace,' I said, 'resisting arrest, assault of an officer.' The sergeant's gaze flickered towards Tiny, who shuffled uncomfortably and nodded.

The sergeant's gaze returned to me. Then back to Jardine. 'Resistance like that would do credit to the French Maquis.' Eyes to me again. 'You know who that is, Brodie?'

'No, sergeant.'

'Nah, I thought not. He's one helluva fucking mess, is all I can say. Used minimal force to restrain him, did you?'

'Yes, sergeant.'

He sighed. 'You know I'm going to have to get the doctor in.'

I returned his sigh and nodded. Medical examinations of injured suspects rarely ended well for the arresting officers.

I got home just before two that morning. Went straight to the cabinet in the bathroom. I was sure I'd brought a bottle of witch hazel from my folks' place when I cleared it out after Dad died. And there it was, behind a bottle of mouthwash and a bunch of prescription painkillers Tiny had taken once for a twisted ankle. I showered and changed and went straight back out. Didn't take long to get over to the east side at that time of the morning.

I was getting a bit fed up by now with the fifteen flights. I was tired after a long shift, and should just have crashed when I got home. But I needed to see her again when I was sure Jardine wouldn't be there. Breathless, I knocked softly at the door. Didn't want to go waking up all the neighbours again. When she didn't respond, I tried the bell and stood waiting in the hall. Not sure why I was so tense, but I was all bunched up inside. Nervous, I guess.

And then the door opened, just a crack, and I saw the curtain of hair hanging down over her face in the dark. I could almost feel her relief as the door opened wider and she stood staring at me with startled rabbit eyes.

'I thought you were him,' she said in a voice so small I could barely hear it.

'He's being detained at His Majesty's pleasure,' I told her.

'What do you want?'

I fished the bottle of clear liquid out of my jacket pocket and held it up. 'I promised to introduce you to Hazel. She's a good pal.'

She looked at the label and a reluctant smile brought light to her dark eyes. She held the door wide and I brushed past her into the sitting room. In the hours since I'd left, she'd made something of an effort to clear the place up. The bottle and broken glass were gone. Chairs righted. There were still feathers everywhere. I turned as she came into the room behind me.

'Got some cotton wool?'

She nodded and went through to the kitchen, returning with soft, coloured balls of cotton wool in a clear plastic bag. I sat down on the settee beside her, soaking a ball with the witch hazel and applying it liberally to the bruising on her face and arms.

In keeping with the rabbit eyes, she sat like one caught in the headlights and just let me do it. A patient and long-suffering creature who has learned through experience that resistance is pointless. I was so focused on what I was doing, I didn't notice at first that although she was facing straight ahead, she had turned her eyes in my direction and was staring at me. It came as something of a shock, and I think I might have blushed.

'What?' I said.

'Did Lee do that to you?'

And I felt my hand go involuntarily to the swelling on my forehead. I nodded.

'Why are you doing this?'

It was a good question. Why was I? I couldn't admit that I fancied her. Cos it wasn't really that. I mean, I fancied lots of birds. But there was something . . . compelling about her. Yeah, that's the word. In some way beyond my control, I had felt compelled to come back. It wasn't a decision I took, or a choice I had made. It was her fault, not mine. But all I said was, 'I hate bullies.'

She smiled sadly. 'So do I.'

'So why do you stay with him?'

She just shrugged. There was a word I'd come across recently. Lassitude. It means kind of lethargic. Lacking energy. That's what she was like. As if the hand that life had dealt her owed everything to fate and nothing to choice. 'It's complicated.'

'Then try and explain it to me.'

'Why?' She turned genuinely puzzled eyes towards me.

'Because . . .' She was asking such simple questions and I was finding them so hard to answer. 'Because I'm concerned.'

Her smile then was dismissive, as if to say that I shouldn't be wasting my time, or my concern. She said, 'He's not always like he was tonight.' Her eyes turned down towards wringing fingers between her knees. 'Not when he's sober. Tomorrow he'll be a different man. You wouldn't recognise him. The place'll be full of flowers, and chocolates. He'll have booked us a table in a nice restaurant somewhere . . .' Her voice trailed away and she cast uncertain eyes towards me, as if fearing I wouldn't believe her. 'He treats me well. Spoils me.'

'Aye, until the next time.'

And she saw her own doubt reflected in my scepticism.

'Listen . . .' I took one of her hands in mine. 'You can get away from him if you want. I can recommend a refuge. There are good people there. You'll be safe. It'll be the first step to a new life. One where being battered by a drunk today isn't the price you pay for being spoiled tomorrow.'

She drew her hand quickly away from mine and wouldn't meet my eye. 'Lee would never let me go. He'd track me down. He'd find me.'

I found myself shaking my head, and knelt down in front of her to take her by the shoulders. It kind of forced her to look at me. 'Mel, as long as I'm around, I'm not going to let him hurt you.'

And I saw such pain then in her black, black eyes. And felt the scorn in the breath that escaped with her words. She shook her head. 'And when you're not around?'

I met Tiny in the locker room when we started our shift at five the following afternoon. He was still in a mood with me, and we shared our own little pool of silence amid the banter of the guys finishing up and the officers just starting. No one seemed to notice. But there was a definite lull in the conversation when Joe Bailley stuck his head round the door and said that the sergeant wanted to see me and Tiny in his office toot sweet.

This was our regular sergeant. Not the duty officer on the charge bar from the night before. Frank Mulgrew was a big man with a ring of fuse-wire ginger hair around his otherwise bald pate.

'Shut the door,' he said when we went in, and we knew then that we were in trouble. He sat behind his desk and glared up at us from beneath bushy ginger brows. He lifted a handful of clipped sheets from the desk and dropped them again. 'Medical report on one Lee Alexander Jardine. Extensive bruising, couple of cracked ribs, concussion, broken nose. Injuries not exactly consistent with a simple case of resisting arrest.'

I said, 'He was drunk, Sarge. Headbutted me and came at Tiny fists flying.'

He cast a sceptical eye over the two of us. 'Is that right?' He lifted the medical report again. 'A couple of big fellas like you needed to inflict this much damage just to restrain a drunk man?' He almost threw it back on to the desk.

Me again. 'He was well gone, Sarge, wouldn't come quietly.'

Mulgrew got slowly to his feet, brown-speckled green eyes bathing us in the light of his contempt. He placed clenched fists on the desk in front of him and leaned forward on his knuckles. 'You are so fucking lucky that Jardine's common-law missus didn't want to press charges against him. And he just wanted out of here so fast he wasn't interested in raising a complaint against you two.'

'You let him go?' I couldn't believe it.

'Maybe you'd have preferred to face disciplinary charges, Brodie.'

Which shut me up.

Mulgrew raised himself up to his full and not inconsiderable height. 'Cross the line one more time and I'll make it my personal mission to see you both out of uniform before you can say Section 38. Now get the fuck out of my sight.'

*

Tiny didn't say a word until we were safely ensconced in the BMW. Even then he just sat silent behind the wheel for the longest time before he turned a look on me that would have wilted flowers. I'd never felt the full force of his fury before. It came in softly spoken words that delivered each blow like a punch.

'You ever do that to me again, Cammie, I'll no' stay silent. I'll fucking shop you. I've worked hard to get where I am. No chance I'm going to throw it all away for some wanker with a hard-on.'

CHAPTER SIX

2051

Earlier Brodie had turned on lights all over the flat, tearing it apart to look for more whisky. Finding, finally, a drawerful of miniatures collected from flights and hotel rooms.

Now he sat in the sad, harsh light of the front room working his way through them, one after the other, as he watched more images of the past slide across the screen above the fireplace.

In the early photographs, when Addie was just a baby, Mel had been happy and radiant, and he lingered over them. But the increasingly haunted face she presented to the world in later years made him scroll more quickly by. Somehow, when someone close to you loses weight, and sadness leaches the life from their eyes, you're not always aware of it. Not at the time. It's only later, and with the benefit of hindsight, that you see it. The before and the after. And it is shocking. Brodie was shocked now that he hadn't seen it in the moment. Or maybe hadn't wanted to. And couldn't face the self-recrimination on top of his self-pity.

He focused instead on Addie. How easy it was in the digital age to take photographs and videos. Hundreds of them, thousands of them. Most languishing on hard drives and SD cards, seldom viewed beyond the taking. But there could hardly have been a generation in history whose lives had been more visually chronicled than Addie's.

Dozens and dozens of her as a baby. Newborn, crusty and wet, fresh from the womb. First nappies, first pram, first cot, first step. Every first of almost every day recorded for posterity.

Mel had told him the night they met that her mother had named her after one of the Spice Girls. Or maybe two of them. Mel herself had wanted to name their daughter after *her* favourite singer. Adele. In the early days she'd always had music playing. The sad, haunting, self-pitying songs of Adele had predominated, to the point where Brodie had come to hate them. Lyrics mirroring a generation obsessed with itself. Though he never said anything.

When Mel told him she wanted to name their daughter after the singer, he had bitten back an objection. There was not a single happiness he would have denied her. But right from the start, he had been unable to bring himself to call her Adele, shortening it instead to Addie. Which had stuck. All her life.

Now he watched through a haze of alcohol as his daughter grew up before his eyes. From laughing toddler, to the solemn-faced five-year-old in her brand-new uniform whose hand he had held as he walked her to the school gates for the very first time. He could remember, still, the sense of loss he'd experienced watching her passing through them and into her new life. The loss was one of innocence. He understood now that each chapter of our lives changes us irrevocably. That we grow and adapt to fit the new narrative. And that nothing is ever the same again.

But he had loved that little girl. And loved her again as he watched her once more grow towards womanhood. The video that Mel had taken of him teaching her to swim. Then the first wobbling turn of the wheels as she learned to ride a bike, screaming, 'Don't let go, Daddy, don't let go,' long after he had.

Now a toothy twelve-year-old with braces, arms wrapped around the astrological telescope he had bought for her birthday. She was almost unable to contain her glee. An obsession with the sky, an early indicator of where the future might lead her. Now she could see the stars that she had somehow always wanted to reach for.

Then came the succession of inappropriate boyfriends that punctuated her teen years, the knee-jerk rejection of parental advice as she very nearly drowned in a sea of adolescent hormones, almost unrecognisable from the little girl he had taken to school that first day.

And finally, the very last photograph he had of her. One she had taken herself. A defiant, accusatory selfie. Her anger at the world – and more specifically, her father – was evident in the curl of her lips, the fire in her eyes. He could barely bring himself to look at it. How had he even acquired it? He had no recollection now. Maybe she had sent it to him. A farewell gift.

Of her hatred and contempt. The force of it had not diminished in all the years since.

He fumbled on the cushion next to him among the empty miniatures, in search of one with an unbroken seal. He found the last one. A bulbous, dented little bottle. Haig Dimple. He tore off the lid and sucked at the neck of it till it was empty. His breathing was stertorous in the still of the room. He closed his eyes and felt the world spinning away. Then opened them to find Addie still directing her hostility at him. He shut his eyes again to escape the pain of it. And made a decision.

CHAPTER SEVEN

It was just over a mile and a half from his tenement home in Gardner Street to police HQ at Pacific Quay. As he did most days, he walked it, avoiding Dumbarton Road where he could. It took him a little more than half an hour. His parka kept him dry in the rain for up to two hours, and his waterproof leggings saved his trousers from a soaking. He wore a baseball cap beneath his hood to keep the rain out of his face.

Today it was falling in a steady, breathless stream, just a degree or two above turning to snow. He hardly noticed it. His head hurt and his mouth was like the bottom of a birdcage. But he didn't much notice that either. His mind was somewhere else altogether, and he was only vaguely aware of the extra weight on his back from the weekend pack chock-full of climbing gear and a change of clothes. He was accustomed to taking it on much more arduous expeditions than this.

The south side of the river was almost obscured by rain as he walked across the Millennium Bridge. The multistorey blocks that housed the media and the police stood wraithlike against a grey sky indistinguishable from the horizon, hazy electric light illuminating misted windows in the gloom. He felt better for the walk. But only just.

DCI Maclaren's door stood ajar, as it always did. Paying lip service to the open-door policy that he had promised but never quite delivered. His familiar bark told Brodie to enter when he knocked on it.

'Got a minute, sir?'

Maclaren looked up. Brodie had changed out of his wet gear, and only his reddened cheeks betrayed evidence of his half-hour walk in the rain. 'What is it?'

'Just wondered if you'd managed to get someone to fly up to Kinlochleven.'

'Aye. McNair's going. He'll be taking a water taxi down to Helensburgh within the hour.' He tugged at his collar to loosen his tie below an outsized Adam's apple that seemed to slide up and down his neck like a gauge recording levels of profanity. 'He's not very pleased about it. The spot where the body was found is halfway up a fucking mountain. The only climbing McNair's done in the last twenty years is into his bed.' He paused for a moment and frowned. 'Why?'

Brodie said, 'I could go.'

Maclaren's frown deepened. 'What happened to your medical condition?' The way he stressed the word *medical* betrayed a certain scepticism.

'I've been given the all-clear, sir. I brought in my backpack and my climbing gear just in case.'

'Well, la-de-fucking-da. McNair will be your friend for life.'

He shuffled through the detritus on his desk to retrieve a buff-coloured folder and held it out. 'The background's all in there. The water taxi will take you downriver to pick up an eVTOL from the temporary airbase at Helensburgh golf course. You'll go via Mull to pick up the pathologist. She's been there carrying out PMs on the victims of the Tobermory fire.' He paused. 'You do know what an eVTOL is, don't you? Never know with you old-timers.'

'It's an electric vertical take-off and landing vehicle, sir. What us *old-timers* used to call a chopper.' He paused long enough for Maclaren to register the sarcasm, then said quickly, 'When was the body actually discovered?'

'Three days ago. Too bloody long. They've been keeping it in some kind of cold cabinet. It was found by a young part-time meteorologist who's married to the local cop. She was up there servicing a mountaintop weather station. Installed it herself apparently, along with a whole bunch of others in the area about six years ago. She now only works a few hours a week, on a service and maintenance basis. Childcare issues, apparently.'

Brodie felt the skin tighten across his face.

'I need you to determine whether it was an accident or foul play and report back.'

Brodie said nothing. He was still reeling.

'If it's foul play, we'll have to send in a full team.' He looked at his watch. 'You'd better hurry. The water taxi's booked for half past.'

CHAPTER EIGHT

Brodie stood on the landing stage. Its raised helipad extended from Pacific Quay out into Cessnock Dock. In the last century they had built ships here, a forest of cranes lining each side of the river, breaking the skyline like dinosaurs. Both species now extinct.

From where he stood, the levees blocked his view of the north side of the Clyde. He huddled down under his parka hood, watching the rain drip from the brim of his baseball cap, a curtain of water obscuring his view of the flooding in Govan Road that extended all the way up to Ibrox Stadium.

He very nearly didn't hear his water taxi coming, its rotors beating almost silently in the rain. Water taxi was something of a misnomer. It was a taxi, certainly, one of many that ferried passengers up and down the river between the city and the temporary airbase at Helensburgh. But they never touched the water. They were smaller versions of the eVTOL that Brodie knew would take him up to Loch Leven. Like grown-up incarnations of the drones he had played with as a kid. Eight rotors in a circle around a glass bubble that carried four passengers.

They flew above the river at something like two hundred feet, keeping to the left of an invisible centre line, obeying the rules of the road as if the river itself was some kind of highway. Which, Brodie supposed, it was. They had a good range, around three hours flying time, and could recharge wirelessly in fifteen minutes on any compatible helipad.

It touched down lightly on the pad in front of him, its coloured navigation lights cut through by rain. A door detached itself from the bubble and slid back as the rotors ceased to turn, and Brodie ran, crouched, through the downpour. A photoelectric cell mounted in the door frame read the card he flashed at it, and Brodie could see his face and ID appear on the driver's screen inside. The driver leaned towards him. 'Alright, pal. Jump in so I can shut the fucking door.'

Brodie swung his backpack through the open doorway and pulled himself up into one of four seats that faced each other behind the driver. As soon as the sensor in the seat detected his weight, a soporific American-accented female voice prompted, *Buckle up, buckle up*, repeatedly until he did. The door slid shut.

'How could they no' get a Scottish wumman to say that? Fucking American cow. That's all I get all day long, mate. Drives me round the fucking bend. Try not to drip all over the good leather, eh?' The driver engaged the rotors, and the eVTOL jerked gently as it lifted away from the ramp and banked across the levee to the river. 'Another beautiful day,' he said, without a hint of irony. He hovered for a moment until another water taxi had passed, before swooping up and out over the water.

Brodie could never quite get used to the lack of engine noise. The cab was cocooned in a silence broken only by the sound of rain on glass. The Clyde lay like some long grey slug beneath them, vanishing into the misted distance. Glasgow itself sprawled away into the rainfall, north and south, sporadic areas of flooding catching and reflecting what little light there was in the sky, like a random patchwork of paddy fields.

'Cop, eh?' the driver said, half glancing back over his shoulder.

Brodie grunted.

'Should be out catching crooks instead of swanning off down the Clyde coast.' He grinned into a rear-view screen. 'Holiday, is it?'

'Aye, right.'

'Seriously, though, crime in Glasgow's beyond a fucking joke these days, know what I mean, mate? Break-ins, carjackings. The lot. You're not even safe in an e-chopper noo. Don't know what the hell the government's playing at. Mind you, who are you gonna vote for? The fucking Ecologists? Gimme a break.'

Brodie was aware of him looking in the rear-view screen again, but pretended that his interest had been drawn by something way below them and off to the north. Why did taxi drivers always assume you were a kindred spirit?

'I blame the immigrants, me. All those . . . what are we supposed to call them now? Asians. Flooding in. Scuse the pun. Hardly ever see a Scottish face these days.'

Brodie couldn't stop himself. 'And what does a Scottish face look like?'

'Like yours, mate.'

'White, you mean?'

'Aye, well, pink in your case. What's wrong with that?'

'There are plenty of Scots who're not white.'

'I'm talking about real Scots, pal.'

'So am I. Folk born here. Second, third generation. As Scottish as you and me.'

The hundreds of thousands of immigrants fleeing climate catastrophe in Africa and Asia had been welcomed in through the Scottish Government's open-doors policy. A policy prompted by concern over falling birth rates and extended life expectancy – an economically unsustainable demographic. But a policy that had fed a growing sense of protectionism, blatantly manifesting itself now as racism. The closed-doors policy pursued by the government in England, on the other hand, had only served to increase clandestine immigration, leading to soaring crime rates there, and even worse discrimination.

The driver said, 'Baw-locks! Just cos they sound Scottish doesn't mean they are.'

'Aye, and just because the words coming out of your mouth bear a passing resemblance to the English language doesn't mean they make any sense.'

The driver cast an aggrieved glance towards the rear-view screen. 'Who fucking rattled your cage?'

Brodie shook his head and averted his gaze to the landscape

drifting by below. Mercifully the driver took the hint and sat in brooding silence for the remainder of the journey.

Large wipers worked overtime to keep the glass free of rain. But it was a losing battle. The world was visible only through sheets of water that constantly distorted it.

Away to the south-west, Brodie saw the inundation surrounding the towns of Renfrew and Paisley. Water that lay in dull reflective sheets shimmered off into more rain. In the early days, the flooding had quickly subsumed the low-lying ground at Abbotsinch, north of Paisley, an area once criss-crossed by the runways of Glasgow Airport. An international hub where hundreds of flights had come and gone each week was now little more than a haven for water birds and fishermen.

They flew over the Erskine Bridge, and as they headed further west, Brodie could just pick out the taller buildings and church spires rising above the floodwaters which had claimed the lower-lying areas of Port Glasgow, and Greenock and Gourock. On the north bank of the estuary, the peninsula of Ardmore was now an island, not much more than a pinnacle of rock. And as they banked to the right, he saw that the snow-peaked mountains to the north were lost in cloud. It was impossible to tell where the land ended and the sky began. Immediately below them, the entire seafront at Helensburgh was gone.

The water taxi swooped over the town and up to the fingers of green that extended across the hilltop. What had previously been the golf course was now a temporary airbase for civilian, and some military, traffic. The extent of its links to the rest of the British Isles was limited by the range of the eVTOLs that served it. International flights were out of the question, except in hops via England, or the recently reunited Ireland, to Europe. Transatlantic flights in and out of Scotland had ceased a long time ago.

The old-fashioned cream clubhouse above the town comprised a jumble of steeply sloped slate roofs, chimneys and dormers, expanding to lounges and a pro shop under several flat-roofed extensions. It stood surrounded by winter-dead trees stripped of their leaves by a series of ice storms the previous month. Taken over by both military and civilian air traffic controllers, it was a hub of airborne activity, with drones and eVTOLs coming and going in a daily traffic halted only by extremes of weather.

The main helipad occupied the former eighteenth green and was surrounded by smaller satellite pads that handled the incoming and outgoing flights of aircraft like the water taxi that had brought Brodie downriver.

The driver settled his e-chopper with a slight bump on the pad furthest from the clubhouse. A much larger eVTOL stood on the main pad awaiting Brodie's arrival. The driver squinted at it through the water streaming across his windscreen. 'That yours, do you think?' It was the first time he'd spoken in about twenty minutes.

'Looks like it.' Brodie struggled into his still-wet backpack.

'Where are you going in that, then?'

'Out to Mull, then Loch Leven.'

The driver turned as the door slid open, and there was something malevolent in his half-smile. 'Rather you than me, mate.' He tapped his screen. 'According to the weather reports, we've got a nasty ice storm incoming this afternoon. Get caught in that, and yon big bird'll drop oot the sky before you can say "ice on the rotors".'

Brodie pulled on his baseball cap. 'Thank you.'

But his sarcasm only amused the driver further. 'Yer welcome, pal.'

Brodie's face was wet and stinging from the cold before he'd taken barely a dozen steps. Icy water seeped in around his neck and his cuffs as he dashed across the neatly manicured grass towards his waiting eVTOL. It stood dripping in the pewtery late morning light. Built more like a conventional aircraft with an extended fuselage, its rotors were mounted at the end of either wing, on forward extensions, and on a V-shaped tail at the rear. Six in total. The aircraft sat on three legs splaying out front and back, and the cabin, like his water taxi, was made almost entirely of smoked glass.

As he reached it, Brodie saw a figure clad in luminous yellow oilskins hurrying towards him from the clubhouse. Old-fashioned cotton cloth waterproofed with oil, he assumed, since plastics had been banned for years now. Brodie stood, dripping impatiently, on the pad. When the technician reached him, he pulled a contactless card reader from under his cape and held it out towards Brodie. 'ID,' he barked through the wet.

Brodie flashed his card at the reader and the technician satisfied himself that this was indeed the police officer whose arrival was expected.

'Cool,' he said, and waved an RFID card at the nearside door of the aircraft to open it, then held it out to Brodie. 'Use this to secure the aircraft at destination.'

Brodie frowned. 'Won't the pilot do that?'

The technician laughed. 'There is no pilot, pal. Well, there is.' He jerked his thumb over his shoulder. 'He's in the club-house. This thing'll fly itself. It's been preprogrammed. Pilot's got a watching brief in case anything goes wrong. We've not lost one yet.'

'I'm not surprised if he's always sitting in the clubhouse.'

Oilskins pulled a face. 'Very funny.' He reached past Brodie and pulled open a flap on the side of the fuselage. 'Retractable charging cable's in there. Two hundred metres of it, which should be more than enough. Get it charging as soon as you arrive. No wireless charging available on the football pitch, I'm afraid.'

'Football pitch?'

'Aye. That's what we've been using as a temporary landing area at Kinlochleven. There's a charging terminal installed at the changing pavilion. And your hotel's right next door. The International.' He nodded towards the interior. 'You'd better get in out of the rain. Sit up front so you've got access to the computer screen.'

Brodie heaved his pack into the rear of the cabin and pulled himself in, sliding across into one of the two front seats. Oilskins climbed in beside him.

'We took out the passenger seats in the back in case you're returning with a body. And the pathologist's usually got a fair amount of gear.' He leaned forward and tapped the middle of the screen twice with his index and middle fingers. It immediately presented a welcome screen. A photograph of the eVTOL taken against a clear blue sky on a bright, sunny day. Brodie thought that it couldn't have been taken here, or at any time recently. A soothing, English-accented female voice introduced herself. *Welcome to your Grogan Industries Mark Five eVTOL air taxi.*

Oilskins said, 'Zebra-Alpha-Kilo-496. Eve, activate remote.'

The screen flickered and displayed an aerial topographical map with their projected route marked in red.

Remote activated, Zak.

Oilskins turned a wry smile towards Brodie and shook his head. 'Even the damn machines call me that now. You been in one of these things before?'

'Never.'

'You'll get the computer's attention just by saying *Eve*. She'll put you in direct communication with the pilot if you have any problems or questions. You can watch a movie if you want, or catch the news.'

Brodie couldn't imagine that he would be doing anything except sitting on tenterhooks until Eve had put him safely on the ground again.

'So, if you've got no questions, I'll let you get on your way.' Zak slipped off his seat to jump down to the pad.

Brodie said, 'I'm told there's an ice storm coming in.' He cast eyes around him. 'I hear these things don't do too well in ice storms.'

Zak grinned. 'No worries, mate. You should reach destination long before she arrives.'

'She?'

'Aye, she's a named storm. Hilda, they're calling her. A German name. Means battle, or war, or something.' He grinned. 'Let's hope you and Eve don't get into a fight with her.' He laughed now. 'Only joking. Eve'll take care of you. They've programmed a bit of a detour, via Glencoe, just in case it gets a bit blowy before Hilda actually gets here. It's more sheltered that way and you'll be able to maintain max speed of about 200k. You'll be fine.' He pressed a button on the inside of the door frame and jumped down as the door slid shut.

Brodie felt himself encased again in silence, save for Eve urging him to *buckle up*. The sound of the rain retreated to a distant patter, although it still streamed down the windscreen. Zak vanished at a run towards the clubhouse and Brodie felt more than heard the rotors starting up. Through the sweep of smoked glass overhead, he could see them rapidly reach speed before Eve lifted gently off the pad, rising slowly into the rain. The rotors canted unexpectedly, angling themselves into a semi-vertical position to provide forward thrust, and the eVTOL shot off suddenly across the roofs of the clubhouse and the trees, lifting higher as it did. Still there was no sound, and Brodie, sitting alone in this strangely alien environment, felt oddly disconnected from the world, as if he had just surrendered his present and his future to some invisible guiding hand over which he had no control.

Eve flew low and fast above the sodden winter ground below. Over the Gare Loch and its long-abandoned nuclear submarine base at Faslane. Loch Long with its lost village of Arrochar, drowned by the storms and the accompanying rise in sea level, cutting off direct access to the West Highlands by road.

Most of the settlements along both shores of Loch Fyne were gone. Strachur, Auchnabreac, and much of Inveraray.

There was snow lying on higher ground now, and the mountain ranges to the north – when you could see them through the cloud – were mostly blanketed by it.

Of course, Brodie knew, Scotland had escaped relatively lightly. Large parts of eastern England had simply vanished under the North Sea. From Hull, as far inland as Goole and Selby. And to the south, Grimsby, Skegness, Boston, King's Lynn. Great Yarmouth and Lowestoft had barely survived. On the west coast, the bright lights of Blackpool had been washed away. Lytham St Annes and Southport were gone.

Much of London was underwater, too. The authorities had moved too slowly in replacing the old Thames barrier, and had run into funding problems when building the levees that would have protected the estuary.

On the near continent, most of the Netherlands, including

Amsterdam and Rotterdam, had been reclaimed by the sea. A good chunk of Belgium, the German seaports of Hamburg and Bremen, as well as large swathes of the western seaboard of Denmark had also succumbed to rising sea levels.

There was worse, much worse, elsewhere in the world. But there was a limit to how much you could absorb before you became waterlogged yourself by too much information. It was one of the reasons Brodie had simply stopped listening to the news, or reading the newspapers, or watching TV. It was depressing beyond words. Suicide rates, he knew, were soaring. Because above all, there was nothing you could do about it. Any of it. So, like many others, he had simply zoned out, limiting consciousness to his own little bubble of existence. The only place he had any say in how things played out.

Now they were flying over the mouth of Loch Linnhe towards the Inner Hebrides, leaving the mainland behind. White caps broke the surface of a turbulent sea below, and Brodie could feel the wind buffeting his eVTOL. Snow started to fall as they reached land again, and Eve banked north-west over the Isle of Mull. Brodie saw how the Atlantic Ocean had nibbled away at a rocky coastline that rose well above sea level in most places, keeping the bulk of the island intact. Tobermory at the north of the island, where his pathologist awaited, had fared less well. The dock and coast road were awash; the row of multi-coloured seafront properties that featured on island postcards was semi-submerged. The rest of the town rose steeply up from the water, cowering among the trees. The coastline of

Calve Island opposite had been completely reconfigured by the relentlessly rising ocean.

Eve lifted up over the town to the golf course that sprawled across the hill to the north of it. There was no helipad here. Brodie saw how they were manoeuvring to land on an almost perfect circle of manicured green. He could even see the hole. Someone had removed the flag, and a figure, huddled in waterproofs and hood, stood in a bunker at the edge of the putting surface with two large slate-grey Storm travel cases. He could barely see anything beyond the figure because of the snow that was driving in now from the west, though it was wet and not yet lying.

He opened the door and felt large wet snowflakes slap into his face. A woman's voice called out from the bunker. 'Well, come and give me a hand, then! I can't carry these on my own.'

Brodie sighed and wondered how she'd got them here. There didn't appear to be anyone else around. He pulled up his hood and braced himself to face the blast, jumping down on to the green and running at a tilt into the wind. There was little visible of the pathologist's face, with her hood crimped tightly around it. Angry dark eyes flashed at him through the snow. 'You're late!' As if somehow he had any control over departure and flying times. 'Take my kit case, it's heavier.' She lifted the other one and ran for the eVTOL. Brodie gasped at the weight of the case containing her kit as he heaved it up out of the bunker and staggered across the green. She was waiting for him by the open door, and together they lifted the two cases to slide into the back of the cabin. He helped her then to climb in and quickly followed, closing the door behind them.

The howl of the wind was instantly extinguished, and a pall of damp silence hung in the air as Brodie slipped into the front seat beside her. She pulled away her hood to reveal jet-black, crinkly hair drawn back from her face and tied at the nape of her neck. Her complexion was a pale brown, her eyes almost as black as Mel's. She had a small, dark brown mole on the right side of her upper lip. Her lips themselves were full and marginally darker than the skin of her face, but touched with red. A handsome woman. In her late thirties, perhaps, or early forties. She glared at Brodie. 'I've been hanging about there in the wind and the snow for nearly half an hour. Ever since they dropped me and told me you'd be here in a few minutes.'

He protested. 'I have no control whatsoever over the timing of this flight.'

But she wasn't letting him off with anything. 'You must have been late arriving for it, then.'

Irritatingly, Eve's relentless voice was urging them to *buckle up*. He could barely think above it. 'For Christ's sake, do what she says and shut her up.'

They both engaged their seat belts and the voice ceased, leaving them once more in silence.

He glared at her, before nodding towards her cases in the back. 'You're welcome, by the way.'

She scowled back at him from beneath dark eyebrows. And then her face creased suddenly into the most disarming smile, and he saw the twinkle of mischief in her eyes. She thrust out her hand. 'Sita Roy. *Dr* Sita Roy, actually. But you can call me Sita.'

He shook her hand and felt the power of the pathologist's grip in muscles developed by the cutting of bone and the prising open of ribcages. 'Cameron Brodie. *Detective Inspector*, actually, but you can call me mister.'

She laughed out loud. 'Yes, sir.' She half turned towards the computer screen below the windshield. 'Eve, we're ready to go.'

Eve responded immediately. *Thank you, Dr Roy. Hold on tight.* And the rotors above them sprang to life.

'You two are acquainted, then,' Brodie said.

She grinned. 'Eve and I have made many a trip together. We're old friends.'

The eVTOL lifted up from the green and wheeled away, back towards the town, rising as it headed south.

Sita said, 'Eve, what's our flight plan?' And the topographical map displayed earlier reappeared, with the route to Kinlochleven outlined again in red. Sita frowned. 'Eve, why are we taking such a circuitous route?'

Incoming ice storm, Dr Roy. It's more sheltered if we approach via Glencoe.

Sita puffed up her cheeks and exhaled through puckered lips. 'And I'd been hoping for a short flight, too. I never travel well in these things at the best of times.'

CHAPTER NINE

They reached the mainland again just north of Oban. Much
of the port town was underwater, its roll-on, roll-off ferry services to the islands long since defunct. Inland then towards
Tyndrum and banking north to Bridge of Orchy. The snow
was still wet, but lying here; the ominous peaks that flanked
the darkly sinister Glencoe with its history of betrayal and
massacre reflected white where light tore through breaks in
the cloud.

'Where are you based?' Brodie asked Sita.

'Queen Elizabeth University Hospital in Glasgow. But they
send me all over Scotland.'

'Yeah, they said you were doing PMs on victims of the hotel
fire in Tobermory. Is there some doubt about the origins of
the fire?'

Sita nodded. 'Your guys think it was an insurance job. If so,
then technically the fatalities are murders.' She sighed deeply,
lips curled in distaste. 'Two children among them. My American
colleagues call burn victims *crispy critters*. I don't share their
sense of humour. There's nothing worse in my book than

performing autopsies on people who have died in a fire. You get used to the perfumes of the autopsy table, but it takes days to get the smell of burned human flesh out of your nostrils.' She canted her head towards the computer screen. 'Mind if I put on the news? I've been out of the loop for a few days.'

He shrugged his acquiescence.

'Eve, play me the news headlines.'

A voice Brodie recognised as a newsreader at SBC said, 'Good afternoon, listeners. Welcome to SBC Radio One. Here are the news headlines. The United Nations reports that the immigration wars raging across North Africa have reached a tipping point. Sheer weight of numbers is overpowering national defences across the continent, from Morocco to Egypt. Tens of thousands are already feared dead in the conflict. Estimates put populations on the move from equatorial Africa and Asia at around two billion, and South European countries are bracing for a fresh flood of migrant boats across the Mediterranean. The United Nations High Commissioner for Refugees, in a statement earlier today, described national immigration policies around the world as unsustainable.'

They played a thirty-second clip of an interview with the High Commissioner herself. She raged at political leaders in Europe and Africa, describing them as *immoral* and *ostrich-like*, accusing them of burying their heads in the sand. 'The problem is simply not going to go away,' she said. 'We have to address it head-on and find solutions. Simply letting people die is no answer.'

'Jesus!' Brodie said. 'Two *billion* people?'

Sita shrugged. 'That's at least how many people live in coastal settlements around the world, and in those equatorial and sub-Saharan African countries made uninhabitable by rising temperatures.'

He shook his head. It made no sense to him. People were dying from heat at the equator and here they were flying into an ice storm.

Sita looked at him quizzically. 'Where have you been, Mr Brodie?'

'I don't listen to the news. It's too depressing.'

'Burying your head in the sand, then.' She sounded unimpressed.

'Ostrich-like.' He echoed the High Commissioner for Refugees.

Sita spluttered her derision. 'Well, of course, that's just not true.'

'What isn't?'

'Ostriches don't bury their heads in the sand. At least, not to hide from reality. They bury their eggs in the sand and stoop to turn them over frequently. They don't hide from danger, they run from it. At up to seventy kilometres an hour. And if forced to fight, they will. An ostrich can kick with a force of a hundred and forty kilos per square centimetre, enough to kill a lion with a single blow.'

Brodie looked at her in astonishment. 'Wow. How do you know all that?'

She shrugged lightly. 'It's sort of a hobby. I've been teaching my kids all about the animals and birds and fish that'll soon be extinct. It's important they know about the world we've destroyed, don't you think?'

'And human beings? Where do they figure on your extinction list?'

'Oh, pretty high up, the way things are going.'

The radio was still playing in the background. Brodie said, 'I don't want to listen to any more of this.' He turned towards the screen. 'Do you mind?'

She shrugged.

'Eve, stop,' he said, and the news broadcast came to an abrupt end. 'Okay, so I'm not an ostrich. I've just got enough problems of my own to deal with.' And he realised, with something of a shock, that for several hours now he hadn't thought once about the death sentence his doctor had handed down to him just yesterday.

'Don't we all?' She didn't sound sympathetic.

He looked at her. 'So are you one of the two billion, then?'

'I would have been. Except I've been here for nearly twenty years. One of the allocation of so-called skilled immigrants allowed in by the Scottish Government. People were welcoming at first. There was already a well-integrated Asian population here anyway. Both my children were born here and consider themselves Scottish. But since the country's been overrun by immigrants, legal and otherwise, hardly anyone sees us as Scottish any more. They just see brown faces and tell us to go home.'

'Why don't you?' He'd asked it before he realised how it sounded.

She scoffed. 'You're no different, are you, Mr Brodie? This *is* my home. And for your information, where I grew up is gone. I guess that's something else you didn't hear on the news you don't listen to. Kolkata, where I was born, where I trained as a doctor, is somewhere under the Bay of Bengal these days. Lift your head, look a little to the north, and you'll see that Bangladesh is gone, too. A whole country. Just not there any more. A bit like Florida. And large tracts of the eastern sea-board of the US.' Frustration escaped her lips in a hiss. 'Just don't get me started on how the world failed to meet its net-zero targets.' She spoke quietly, but there was a dangerous anger seething behind her words.

Brodie said, 'I thought India was one of the worst offenders.'

She flashed him a look that quickly turned to embarrass-ment. 'It was. Along with China and the US.'

They were following the line of the road now as it wound its way through Glencoe. Jagged peaks on either side pushed themselves up into a broken sky, patches of watery late after-noon sunlight slanting through to land on snow in dazzling patches that came and went like random searchlights in a war zone. The snow had stopped falling, but the wind had risen, and they felt the eVTOL stabilising itself against the buffeting. Here and there, clutches of pine trees pushed themselves up above the snow on some of the lower slopes. The wind blew fresh snowfall off sheer rock ridges in fine clouds that caught sporadic glimpses of sunlight from the sunset beyond the mountains to the west. Tiny, unexpected rainbows appeared and vanished in the blink of an eye.

By the time they reached Glencoe village at the western end of the valley, the sky had darkened, the sun sliding down beyond the horizon, its light snuffed out by the sudden fall of night. The first hail carried on the edge of the wind crackled against the glass.

'What the hell's that?' Sita said suddenly, peering forward into the gloom.

Ahead of them, on the far side of Loch Leven, a phalanx of lights reflecting in the black of the loch blazed around what looked almost like a small city.

'Ballachulish A,' Brodie said. 'A fucking eyesore. Excuse my French.'

'The nuclear power station?'

He nodded. 'This was a beautiful, unspoiled part of the world before they built that monstrosity in the thirties. They generate 3500 gigawatts a year there. Enough to supply electricity to every household in Scotland, they said. I used to come climbing here a lot, and hillwalking. It ruined almost every view, from every hilltop and every mountain.'

'But zero emissions,' she said dryly.

'Aye, zero fucking emissions.'

She leaned forward to get a better look as Eve banked to

the north-east. 'They built it right on the water's edge. Isn't it in danger of flooding?'

He shook his head. 'They demolished the Ballachulish Bridge, just to the west there.' He pointed towards a thin line of red lights spanning and reflecting in the loch. 'Replaced it with a barrier to contain rising sea levels. Generates tidal electricity, too. And they ran a road across the top of it, so you don't have to go round the loch to get to the other side.'

'Ballachulish A,' she said thoughtfully. 'Does that mean they're planning a Ballachulish B?'

He scoffed. 'Probably.'

She said, 'I can see the point of building a place like this somewhere out of the way, but how do they get the spent plutonium out? I can't imagine that it would be very safe by road. Or sea.'

'They don't,' Brodie said. 'They drilled into the bedrock next to the plant. Half a kilometre down, something like that. Then excavated a network of tunnels. That's where they put the waste. Buried for eternity, they say.'

'Eternity, eh? That's a long time. I wonder how they measure it against something that's got a radioactive half-life of 24,000 years.'

Brodie smiled sadly in the dark. 'Well, we'll not be around to find out.'

And she said quietly, 'I wonder if anyone will.'

Eve shook suddenly, as if something had slammed into her.

Involuntarily Sita reached forward to grab the dash. 'What the hell was that?'

Brodie's heart was pounding. 'The wind, I guess.'

Rain turned increasingly to sleet and hail and slashed through the lights of their eVTOL as it followed the course of this Scottish fjord, mountains rising steeply in the dark on either side of a loch that was in reality a glaciated valley flooded by seawater. And all that Brodie could hear in his head were the words of his taxi driver earlier in the day, when he told him there was an incoming ice storm. *Get caught in that, and yon big bird'll drop oot the sky before you can say 'ice on the rotors'.*

Almost as if she had read his mind, Eve reduced her height. Dropping fifty feet or more in a matter of seconds. Sita cried out, 'Woah!' She grabbed Brodie's arm.

They felt the storm snapping at their heels, the eVTOL's stabilisers working overtime. To their right, at Caolasnacon, they saw the lights of houses lining the water's edge and extending back into the trees.

Sita said, 'Is that Kinlochleven?'

Brodie shook his head. 'Nah. These were homes built for workers at the nuclear plant. All 3D printed. Ugly things.'

He saw Sita turning to look at him in the reflected light of the computer screen. 'They're printing houses this far north now?'

He shrugged. 'Apparently.'

In the distance, at the head of the loch, they saw the lights of Kinlochleven for the first time. They seemed feeble somehow, almost smothered by the dark. Arcs of street lamps delineated rows of houses built around the curve of the valley and dissected by the dark passage of the River Leven as it tumbled down from the mountains above, cutting a swathe of fresh icy water into the warmer, salinated seawater that washed up on the shore.

'Tell me that's the village,' Sita said, a hint of quiet desperation in her voice.

'It is,' Brodie said, with his own sense of relief. In just a few minutes they would be on the ground, and the storm could do its worst.

Then all the lights went out, and it was as if they had been sucked into a black hole. There was no light anywhere, except for the reflected glow of the computer screen in the eVTOL, and a couple of spots directed by Eve on to the water below. Water that seemed perilously close now and rushing past at speed.

Sita let out a tiny scream. 'What's happened?'

'Must be a power cut.'

'Power cut?' She almost shouted it. 'You just told me they're generating three and a half thousand gigawatts of electricity at the other end of the loch. How can there be a power cut?'

'I don't know. The storm must have brought down power lines.'

'Well, how are we going to land in the dark?'

'I have no idea!' He turned towards the screen. 'Eve, how are we going to land in the dark?'

Eve sounded unruffled. *I am preprogrammed for landing, Detective Inspector Brodie. I do not require light.*

Sita was still grasping his forearm, her pathologist's grip almost cutting off the blood supply to his hand. Below, by the light of the eVTOL, they saw a shoreline exposed by low tide, white-topped surges washing over it, forced up loch by the wind that scoured the valley, winter-bare trees now bucking and bending among a scattering of houses. The River Leven in spate, white water generated by its power, almost glowing as it fought against the storm surge to feed itself into the loch.

Then they banked against the force of the wind, frozen rain hammering on the glass, and Brodie just prayed that the rotors wouldn't ice up before they landed. A perfectly delineated rectangle of unbroken snow swung crazily into their field of vision. The football pitch. And both of them held their breath as Eve dropped from the air to settle with a bump in the snow, light from her spots spilling out around them in a wide circle that faded into the darkness beyond. The rotors powered down and Brodie felt Sita's grip on his arm relax as they both drew deep breaths.

'Well, that was fun,' she said.

'Not.'

She turned towards him and laughed with relief. 'Nothing like a near-death experience for bringing folk together. Thanks for the use of your arm. You might need witch hazel for the bruising.'

And his heart leapt. A thousand memories of Mel flashing through his mind in a moment. He rubbed his arm to get the circulation back in his hand. 'With a grip like that, it might need to be set in plaster.'

She laughed again. 'Plaster? You're giving away your age now, Mr Brodie. They haven't used plaster to splint broken bones since the Dark Ages.'

They were interrupted by Eve, whose oleaginous tones seemed to coat the interior glass of the eVTOL. Her lack of panic in the landing had been reassuring somehow, although they both knew that panic was not programmed into her software. *Warning. Low battery. Low battery.*

'Jesus,' Brodie said. 'Now she tells us!'

Extinguishing lights to save power. Please connect me to a power source as soon as possible.

Her lights went out, and the computer screen powered down, leaving them in a darkness so thick it felt almost tangible.

'Fuck!' It was Sita's voice that reverberated around the cabin. 'What do we do now?' She heard Brodie swinging himself out of his seat to clamber into the space behind them, and cursing as he stumbled over one of her Storm cases. A moment later, light filled the interior.

'I always keep an LED headlight in my pack,' he said, and she turned to see him stretching the elasticated band around his head so that the tiny lamp projected from his forehead to light the way in front of him.

'Very practical,' she said. 'Can you teleport us to our hotel now?'

'If only I knew where it was.'

'Huh! And just when I was starting to like you.'

'The technician at Helensburgh said it was right beside the football pitch. So it can't be far.'

'Well, I'll let you go and find it. And when you do, you can come back and give me a hand with my stuff.'

'Yes, miss. Whatever you say, miss.' Brodie raised a hand to tug an imaginary forelock.

She grinned. 'Well, it doesn't make sense for both of us to go stumbling about in the dark. And you're the one with the light.'

Brodie pulled a face. 'So I am.'

He pushed the button to open the door, unprepared for the blast of wind and sleet that nearly took him off his feet. He zipped his parka up to the neck and pulled the hood over his head, bracing himself with a hand either side of the door frame before jumping down into the snow. Sita reached over to shut the door quickly behind him.

He turned in a quick arc, but the light from his headlamp didn't penetrate far through the freezing rain that drove into his face. He had a sense that since the wind was coming from the west, that was the direction he should take. So he staggered into it, semi-blind, until he reached a perimeter fence. It stood around two-and-a-half metres high. His face was stinging, almost numb now. There had to be a gate somewhere. He worked his way along the fence. He could see there were trees on the far side of it, and beyond them, a strangely eerie glow that seemed to flicker through branches that creaked and swayed in the wind. And, finally, a gate. It opened on to an area that felt firmer underfoot beneath the snow. Tarmac, perhaps. He leaned forward into the wind and pushed himself up a short slope, where a path appeared, cutting a way through the trees. And there, conveniently planted in the ground, was a white arrow sign. *International Hotel*, it read.

He found his way back to the eVTOL by following his own footsteps, accelerated on his return by the wind behind him. He banged on the glass with the flat of his hand until Sita opened the door. She jumped down into the storm, a tiny circle of wide-eyed face peering out from her hood.

'You found it?' She had to shout above the roar of the wind.

'Yes. But we'll never be able to carry both your cases.' He reached in for his pack and swung it over his shoulders. 'You can get your kit in the morning. You won't be doing any post-mortems tonight.'

'What about recharging Eve?'

'Even if there was any electricity, I haven't the first fucking idea where the charging hub is.' He felt the wind whipping the words from his mouth as he shouted them into the night. 'Nobody's going to steal her tonight.'

She nodded, and reached in to pull her personal Storm case from the hold. 'We can take an end each.'

Brodie grimaced into the rain. 'Of course we can.'

She grinned. 'I always knew a policeman would come in handy someday.'

Brodie smiled, and realised that for someone close to death, he hadn't felt this alive in years. He leaned in to hit the close button and pulled back as the door slid shut.

'I hope you know how to get back into that thing.'

He patted his pocket. 'Got the keycard right here.'

A smile twinkled in her black eyes.

They stooped to take a handle each and lifted her trunk, and set off by the light of his headlamp to follow his footsteps back to the gate.

By the time they had cleared the trees beyond the fence, the International Hotel came in range of their light, a sprawling, cream-painted building on two storeys with a faux tower and pointed dormers. All its windows simmered in darkness, but beyond the glass around the entrance porch at the foot of the tower, a faint flickering light offered the hope that they weren't the only humans still alive in this storm.

They struggled up the half-dozen steps to the entrance, the wind catching and swinging Sita's Storm case between them, and pushed gratefully through the door into a long, tartan-carpeted entrance hall. Candles burned in a reception hatch below a set of antlers, and on a table opposite. The door swung shut behind them, and the storm receded into the night, leaving flames flickering in the hallway to send their shadows dancing around the walls.

Brodie and Sita set her case down and stood dripping on the carpet. There was a residual warmth in here, but it still felt chilly, the air laced by a faint smell of damp. Brodie stepped up to the reception hatch. Glass windows were slid shut and it was impenetrably dark beyond them. A bell sat on the counter and he banged it several times with the palm of his hand. Its shrill ring resounded around the emptiness of the place. 'Hello,' he called into the silence that followed. 'Anyone home?'

Sita said, 'I feel like I've just walked on to the set of *The Shining*.'

'Don't say that,' Brodie said. 'I've never been able to watch that movie beyond the twins in the corridor.'

'Big brave man like you?'

He grunted. 'We all have our demons.'

A door at the far end of the hall swung open, startling them, and the silhouette of a large man approached in the gloom. A candle set in a holder in his left hand cut an oblique penumbra on a bearded face, the larger shadow cast by his bald head and shoulders increasing in size on the wall behind him as he drew nearer.

He broke into a grin. 'Welcome, welcome. You made it, then?' And he laughed. 'Well, of course you did. You're here. To be honest, I wasn't really expecting you, with the storm and all. And there's no telephone, no internet, so how could anyone let me know?' He stopped for a breath and held out a bony hand. 'Mr Brodie? Mike Brannan. I own the place, for my sins.'

Brodie shook it reluctantly, and resisted the temptation to wipe his palm on the seat of his trousers. Brannan turned to Sita.

'And Dr Roy, I presume.'

Brodie stifled amusement at the brief flicker of pain that registered in Brannan's face as the pathologist shook his hand.

'Can't feed you, I'm afraid. No power. Kitchen's out of action.'

Brodie said, 'Alcohol will do.' He glanced at Sita for affirmation. She nodded.

'Yes, please.'

'That can be arranged.' He waved a hand towards the entrance to the Bothy Bar. 'You'll have the place to yourselves. There's not another soul in the hotel. I'll light a fire, if you like. It's a wood burner, so carbon-neutral.' He smiled, as if waiting for a round of applause. When none came, he said, 'I'll show you to your rooms.'

They followed him up the staircase to a long, carpeted hallway with rooms along each side. Brodie had extinguished his headlight to save the battery, and the place felt oddly disconnected from reality.

Brannan half turned a salacious smile towards them. 'Not sharing, I take it?'

'No,' Sita said firmly.

'Thought not.' He opened a door. 'You're in here.'

Brodie and Sita struggled in with her Storm case and heaved it on to a luggage stand. The cream room had purple carpet and curtains, and fresh towels folded on the bed.

'And you're right next door,' he told Brodie. He began lighting candles on the dresser. Clearly power cuts were not an uncommon phenomenon.

Brodie slipped his pack from his back. 'What happened to

Charles Younger's car?' He'd spent some of the flight to Mull reading over the notes that Maclaren had given him. There had been no mention of a car.

Brannan seemed perplexed. 'I don't understand.'

'Younger's car. He must have parked it here at the hotel.'

'Oh, I've no idea. We don't reserve parking places for guests. We were busy last August, so he'd have had to take pot luck. If he had a car at all, that is.'

'How else would he have got here?'

'Haven't the foggiest.'

'But his personal belongings were still in his room?'

'Yes, but they couldn't stay there after he'd gone missing. The room was booked by someone else. So Robbie came to bag it all up and take it away.'

'Robbie?' Sita said.

'Yes, the local bobby.' He chuckled. 'Robbie the bobby. Robert Sinclair.'

Brodie said, 'I used to come here years ago, climbing and hillwalking. There was no local bobby then. The old police station was an Airbnb.'

'Ah, yes,' Brannan said. 'But there was a mini population explosion in the thirties while they were building the power plant, and apparently it was decided to reinstate the local policeman. The old police station was for sale at the time, so they reacquired it and Robbie's your bobby.' He handed Brodie a keycard. 'Here's your key. I'll light some candles in your room and go down to get the fire going.' He paused in the doorway, turning, as if struck by an afterthought. 'Do you want to see the body?'

After a moment's shocked silence, Sita said, 'It's here? In the hotel?'

Brannan shrugged philosophically. 'Well, they'd nowhere else to put him. And I had a big cold cabinet for cakes and desserts lying empty in the kitchen.'

The kitchen was at the back of the hotel, pots and pans and cooking utensils hanging from a metal rack above a central stainless steel worktop. The place smelled of stale oil, taking Brodie back to pub meals in Highland villages and scampi in a basket. The shadows from Brannan's candle cavorted among the appliances and the big overhead extractor units. 'Through here,' he said, and Sita and Brodie followed him into an anteroom that might have served as a pantry. The air was heavy with the astringent stench of detergent.

The cake cabinet stood on castors and was pushed up against one wall. Its glass top was misted so that it was impossible to see inside. Brannan handed his candle to Brodie.

'Here, take this.' And he lifted the lid.

Charles Younger was a man in his forties, big built. Thinning fair hair lay slicked across his forehead. He was still fully dressed, just as he had been found. Vomit-green parka, black ski pants, cheap walking boots. His woolly hat had been recovered separately and lay beside him. He was folded, knees drawn up, to fit into the cabinet. His eyes were open, his mouth gaping, his face bruised and grazed. Those parts of his skin that were visible had taken on a pink-reddish hue.

Brodie was struck by the ice-blue of eyes that seemed to match the colour of his lips. There hadn't been much about him in Maclaren's dossier. A single man. No relatives apart from a very elderly mother who was living in a care home in Livingston. He'd been with the *Herald* since graduating from Edinburgh University. Won numerous awards, and struck the fear of God into any politician who learned that he was digging into their history. Brodie had never read a word he'd written.

'Looks fresh,' was all he said. 'For someone who's been dead for three months.'

'Being frozen in ice most of that time will have preserved him pretty well,' Sita said. 'And this cabinet's what? Three, four degrees?'

Brannan said, 'Usually around four or five.'

'Which means he probably hasn't completely defrosted on the inside yet. Though this power cut is going to accelerate decomposition. Even so, I'm going to have cold hands when I go pawing about his interior tomorrow.'

Brannan lowered the lid on the sightless body inside. 'What I want to know is who's going to pay for a new chiller. I mean, is it an insurance job, or do the cops cough up? Cos, let's face it, no one's going to want a slice of chocolate cream gateau from this one now.'

CHAPTER TEN

The bay windows in the bar rose from a wooden floor to a stucco ceiling and opened, in summer, on to a terrace with unrestricted views back down the loch. There was no view now, though. Just black beyond glass that ran with rain, distorting their reflections. Despite the double glazing, the flames of their candles ducked and dived in the draught, and Brodie watched the glass bend with the force of the wind. He shivered, despite the comparative warmth that came from the fire that Brannan had lit.

A pool table lurked in the darkness of one corner, the balls of a half-finished game casting shadows on the baize. In a flicker of candlelight at the bar, Brannan placed a bottle, a jug of water and two glasses on a tray, threw on a couple of packs of crisps, and crossed to the window. He set the tray down on their table and straightened up, running a large hand back over the shining baldness of his head.

'Shame you can't see the view. It's one of the big selling points of this place. But never mind, you'll see it tomorrow. The forecast's quite good, and you'll no doubt want a drink after . . .' he hesitated and rephrased, 'before you leave.' His smile was unctuous. 'As for tonight, just help yourself to the bottle. I'll put it on your room, shall I, Mr Brodie? No doubt Police Scotland will be paying for it.'

'No doubt.' Brodie grunted and leaned forward to break the seal and uncork a bottle of Balvenie DoubleWood, pouring generous measures of its pale amber into each of the glasses. 'Thank you, Mr Brannan.' It was clear, he thought, that he and Sita wanted some privacy, but Brannan wasn't taking the hint. Or maybe he was just lonely.

'Ironic, isn't it?' he said. 'A nuclear power plant at one end of the loch, and a hydroelectric power station at the other, and all we seem to get all winter these days is power cuts.'

'And why's that?' Sita asked him.

'Because the power still leaves here on pylons. They never invested in underground cabling. So the overhead cables are exposed to the full force of the weather. All these storms. Ice forms on them and the weight of it brings them down. Bloody short-sighted, if you ask me.'

Nobody was, Brodie thought. But kept his own counsel.

'You know, they've had hydro power here since the first decade of the twentieth century. Way ahead of its time. They built it to power an aluminium smelter across the river there. That's long gone now, mind you, but Kinlochleven was the first village in the world to have electricity in every home. The electric village, they called it.' He chuckled to himself. 'When I bought this place six months ago, it was called the MacDonald

Hotel. I toyed with the idea of changing it to the Electric Hotel. But people thought that was a bit shocking.' He laughed. And when neither Brodie nor Sita joined him, he added lamely, 'So I settled for the International instead.'

Brodie took a long pull at his whisky and closed his eyes, trying to shut out the voice, hoping that it might be drowned by the wind. A forlorn hope.

'Wish I'd bought it back in the thirties when they were building Ballachulish A. There was an influx of thousands of workers then, a lot of foreign experts among them. They all needed accommodation. So the International, or the MacDonald as it was then, and every other hotel and B & B for miles around was full. The bars and restaurants were stowed out, summer and winter, for more than five years. Even when they finished work, the plant itself employed nearly two thousand folk, and until they built the 3D homes across the loch, they all needed accommodation.' A long, sad sigh escaped his lips. 'Different story now, though. Business has dropped right off. We still do well in the summer, but the winter's dead. Just dead.'

'Like Mr Younger,' Brodie said.

Brannan leaned in a little, and his voice became softly suffused with a sense of confidentiality. 'You won't be advertising the fact that he was staying here, will you? It wouldn't be good for business.'

Brodie opened his eyes and felt a wave of fatigue wash over him, as if he had just endured a long, sleepless night. 'I'm afraid I can't say what the press will or won't report in relation to the case, Mr Brannan. I suspect that if his death was the result of natural causes, or an accident, they won't pay it very much attention at all.'

'Well, what else would it be?' Brannan seemed surprised.

'Until Dr Roy has conducted her post-mortem, nothing can be ruled out, including foul play.'

The hotel proprietor frowned. 'Murder, you mean?'

Brodie shrugged. He had assumed that this was self-evident.

'But who would want to murder him?'

'We don't know that anyone did. But if he was, then it'll be my job to find out who killed him and why.'

Brannan stood staring forlornly at his reflection in the window. 'Never even thought of that. Let's just hope he fell, or had a heart attack or something. Can't afford to lose any more business.' He folded his arms across his chest.

Sita said, 'With all the snow you get here, you'd think it would be good for winter skiing.'

'Oh, we have the snow, but not the infrastructure. And too much snow, if the experts are to be believed. Ballachulish A might have brought a lot of business, but it also buried us in bloody snowfall.'

Brodie frowned. 'How's that?'

'So, to cool the reactor they use water from the loch, which then goes back in to recirculate. That raises the overall temperature of the loch, making it warmer in winter than the air. So winter precipitation almost always falls as snow. Kind of like the lake-effect snow they get in North America. The stuff just dumps on us. Metres of it at a time.' That thought seemed to draw the curtain on his desire to talk to them any further. He said, 'I'm afraid it'll be a cold breakfast, unless the power comes back on again overnight.' He made a tiny bow. 'Sleep well.' And he retreated into the dark of the hotel from which he had emerged half an hour earlier.

Brodie let out a long sigh of relief. 'I thought he'd never go.'

'Interesting, though,' Sita said, 'that it never occurred to him that Mr Younger might have been murdered.'

Brodie took a thoughtful sip of his DoubleWood. 'Well, in truth, it does seem unlikely. I mean, if someone had killed him, they would hardly drag him halfway up a mountain to get rid of the body.'

'Maybe they killed him up there.'

'Well, there is that. But, then, you'd have to figure it would have been easier to kill the man before he went up.'

Sita emptied her glass and poured herself another. 'You?' She waved the bottle in his direction, and when he nodded, refilled his glass. 'What was he doing up the mountain anyway?'

'Hillwalking, apparently.'

'Ah. A passion, was it?'

'That's the odd thing. He was supposed to be on a hillwalking holiday, but from all accounts he'd never been hillwalking in his life.'

'How did he manage to climb a mountain, then?'

Brodie sucked in more whisky. 'Binnein Mòr's not a difficult climb. Anyone could walk it, really. Take the long way round, in good weather, and in August, and you wouldn't need much experience to reach the summit.' He paused and ran the rim of his glass thoughtfully back and forth along his lower lip. 'But the body was found in a north-facing corrie. Coire an dà Loch.'

'Which means?'

'Corrie of the Two Lochans. And you wouldn't venture up that way unless you had considerable experience.'

They became aware for the first time that the wind outside seemed to have dropped. The rain was no longer hammering against the window. Brodie used a hand to shade his view through the glass from his own reflection and peered out into the dark.

'It's snowing,' he said. 'Quite heavily.'

'Will that make it more difficult for you tomorrow, then, if you're going to go up there to take a look at where the body was found?'

He nodded. 'It will. But I came equipped for it.' He grinned at her. 'And my kit doesn't weigh nearly as much as yours.'

She shrugged. 'Tools of the trade. You don't cut open another human being without the right equipment.' She drained her second glass and refilled it, before pushing the bottle towards Brodie.

He grasped it to pour another. 'And what drew you to doing that?' he said.

'Oh, it was never my ambition to become a pathologist. I wanted to be a doctor, Mr Brodie.'

'Cameron,' he corrected her. But she just smiled.

'I trained at the Medical College and Hospital in Kolkata for five years to get my MBBS.'

'Which is what?'

'Bachelor of Medicine, Bachelor of Surgery. We had a guest lecturer in my fifth year, a visiting American pathologist, and when he took us step by step through an autopsy, I was intrigued by just how much you could tell about a person from their dead body. How they had lived. How they had died. And I was struck by something he said. He told us that when he performed an autopsy on the body of a murdered person, he felt like their last remaining representative on this earth. The only one able to tell their story, explain how they had died, even catch their killer.' She smiled. 'And that's when I decided I wanted to be a pathologist.' She issued a self-deprecating little laugh. 'Maybe I'd have thought differently about it at the time if I'd realised it would involve another four years of specialty training.'

Brodie was amazed. 'Nine years' training to cut open dead bodies. But just five to make folk well again?'

She laughed. 'Yes. Seems like it should be the other way round, doesn't it? But I enjoyed my time there. The Kolkata Medical College was the second oldest in Asia to teach Western medicine. And the first to teach it in the English language.' She raised a hand to pre-empt his comment. 'And before you say anything, I know my English is good. In my opinion, I speak it better than most Scots.'

He chuckled. 'That wouldn't be difficult.'

She was getting through her Balvenie DoubleWood at a good lick, and there was a glassy quality now in her eyes. 'So what else should I know about Mr Younger before I go cutting him up tomorrow?'

Brodie shrugged. 'I don't know that much myself. An investigative journalist with the *Scottish Herald*. Single. *Not* a hillwalker, despite the reason he gave people for being here. It was Brannan . . .' he nodded vaguely towards the interior of the hotel, 'who reported him missing when he didn't return to check out and pick up his belongings. There was no real search for him, because nobody knew where he had gone, where to look.' He swirled some whisky pensively around his mouth. 'One thing, though. There's about a minute or so of CCTV footage of him on the day he disappeared. Talking to someone in the village. A man, apparently, who has never been identified.'

'You've seen it?'

He shook his head. 'No. But I should be able to view it at the local police station. They record all the feeds there from around the village.' He lifted the bottle and held it up against the candlelight. They were about two-thirds of the way through it. He raised an eyebrow in admiration. 'You can drink,' he said.

She raised her glass. 'So can you.'

He laughed. 'Goes with the territory, I guess. Folk like you and me, we see things that most people never do. When I was a traffic cop, I lost count of the number of times I attended road accidents where we had to cut people out of their cars in pieces. Or as a detective investigating a murder where the victim had been hacked to bits. Most murders aren't pat and clever constructs like they write about in books. They're just brutal and bloody.' He paused. 'Well, you'd know all about that.'

She nodded. And it was his turn to refill her glass before topping up his own.

'So . . . you mentioned kids earlier. You're married, I take it?'

'Was.'

'Oh. Divorced?'

'Widowed.'

And for the first time he saw a sadness behind her eyes, and realised it had always been there. He just hadn't noticed before.

She took a gulp of whisky and held it in her mouth for a long time before finally swallowing it. 'Viraj. We were at school together. A lovely boy. Fell head over heels the first time I ever set eyes on him. He had such big eyes, and luscious curls that fell about his forehead. I could only have been about eight.' She smiled sadly, replaying some fond memory behind the increasing opacity of her eyes. 'I went to medical school, he trained as a computer programmer. We were sort of an item off and on for years. Then, when I came to Scotland, he followed me here. Got a job in what they laughingly called Silicon Glen, and told me he wasn't about to let me escape that easily.' She laughed. 'What's a girl to do? When a man demonstrates his love like that, and gets down on one knee to propose . . .'

She stared into her glass now, as if the amber in it provided some window to the past.

'We had two beautiful children together. Palash. Two years older than his little sister, Deepa. They're nine and eleven now.' She looked up over her glass at Brodie. 'My whole world.' And he wondered how much of this she would be telling him if it wasn't for the whisky.

'What happened?' And he knew it was the whisky that emboldened him to ask. But he did want to know.

'I was working one night at the mortuary at the Queen Elizabeth Hospital. I'd just called home, expecting him to pick up. I left a message, then called his mobile, but there was no answer. I knew he'd been out earlier, and the kids were over-nighting with school friends. I just wanted to say I was going to be late that night. They'd just wheeled in a body. Victim of a street attack, and I had to do the PM.' The deep breath she drew had a tremble in it as she tried to control her emotions. 'I went to the autopsy room to open up the body bag. And there was Viraj, lying there staring back at me from the slab. My beautiful boy with his big brown eyes, and those gorgeous curls falling over his forehead. Sticky with blood now. His face all swollen and broken. Missing teeth. Beautiful white, even teeth he'd had. Lips all split and bloody. Lips that had kissed me so many times. A random attack, they said. Kids whipped up into a racist fury by anti-immigration politicians. Killed for the colour of his skin.' Her voice cracked. 'Dead because he followed me here.'

A silent tear tracked its way from her eye to the corner of her mouth.

Brodie was shocked to his core. 'I can't imagine.' His voice was the merest whisper in the dark.

'No, you can't,' she said, as if daring him to even try.

He had no idea what to do, or say. And they sat in silence for the longest time. Until finally she drew a long, quivering breath and wiped away the tear. She took a sip of whisky and cleared her throat, a determined effort to change the direction of their conversation.

'So what about you?'

'What about me?'

'Married?'

His eyes dropped to the glass he clutched in both hands. 'Widowed,' he said, and he felt her eyes on him in the dark.

There was another long silence before she said, quietly, 'Do you want to tell me?'

He closed his eyes and thought that probably he didn't. He had spent most of the last ten years trying to forget. Images burned into his retinas, scorched into his memory. Pain that had never left him in all that time. And yet, hadn't Sita just bared her soul to him? The whisky speaking, for certain. But she had told him things she had quite possibly never revealed to anyone. Opened up her own little box of horrors to public view. How could he refuse to reveal his to her? A grown-up version of 'you show me yours and I'll show you mine'.

As if she could read his mind, she said, 'It's okay, you don't have to.'

But he wanted to now. As if some invisible constraint had suddenly been removed. He *needed* to share it with her. Things he had never spoken about to anyone. And with the sense of his own death little more than a breath away, he felt the urge almost to shout it from the rooftops.

'I was on the night shift,' he said. Then looked up. 'Why is it these things always seem to happen at night?' He remembered it had been a warm, humid Glasgow night. He'd had a fish supper earlier, liberally sprinkled with salt and vinegar. And he still recalled the taste of it in his mouth when he threw it up just a few hours later. 'I was a detective constable then. Working out of Pacific Quay. I got a message that the DS wanted to see me. I thought he was behaving kind of strange. Told me that he was taking me off shift. That I was needed at home. Said he didn't have any further information. But I could tell that he did, and I knew that something awful must have happened.'

His recollection of it was painfully vivid. The frantic drive across the city. Turning into the road where he lived. The two police cars, and an ambulance, sitting outside his home. Neighbours standing at gates, gazing from windows, an intermittent blue cast on inquisitive faces.

'I ran up the steps to the door. There was a cop in uniform barring the way. He raised a hand and asked where I thought I was going.'

He heard himself shouting. *It's my fucking house!*

'Someone was crying inside. My daughter. Just crying and crying. Throaty, like she had cried herself hoarse. Which she had.'

Sita sat perfectly still. 'What age was she?'

'She'd have been seventeen then. Just started at Glasgow Uni. Everyone was upstairs. A cop on the half landing, and a couple of ambulance men a few steps above him. Addie was sitting on the bed in our room, a policewoman with an arm around her. She was inconsolable. There was a medic. A woman. She was standing in the open door to the bathroom. I still remember her turning towards me, eyes wide with shock, face the colour of chalk. And she must have seen things in her time.'

He paused to draw breath. Closing his eyes and replaying it all in the dark.

When he opened them again, he said, 'She advised me that it would be better to remain on the landing. Like there was a chance in hell I was going to stand out there. I glanced into the bedroom and Addie was staring back at me. The look on her face . . . I . . . I've seen it every night since, when I'm trying to sleep. The accusation in it. The naked hatred. I felt, right there and then, like my life was over, whatever it was that lay beyond the bathroom door. But still I had to look.'

He turned his head slowly towards the window, as if it might offer a reflective insight into the moment. Wet snow slapped the black pane and ran down it in slow rivulets, like tears.

'I pushed past the medic and stepped into the bathroom. The overheadlight seemed unnaturally bright, reflecting back at me off every tiled surface. Like some overexposed film.' He shook his head. 'Of course, I realise now it was just my pupils that were so dilated with the shock.' A series of short, rapid breaths tugged at his chest. 'Mel was lying naked in the bath. Her eyes were shut, and there was this strange, sad smile on her lips. First time I'd seen her smiling in months.' He turned away suddenly from the window, as if he could no longer bear the vision it was offering him. 'The water was crimson with her blood. Marbled darker by it in places. The woman I'd loved since the first time I ever set eyes on her was dead.'

He turned now towards Sita.

'Took her own life. It was Addie who found her. Came home from a night at the student union, and . . .' He couldn't bring himself to finish. 'I'd give anything to be able to erase that moment from her life. It's when she stopped being my little girl. It's when she started hating me.'

Sita's brows crinkled into a frown. 'Why would she do that?'

'Because she blamed me. Mel left a note, you see.' He gave a sad little chuckle that nearly choked him. 'She wasn't the most . . . literate person in the world. Articulate in every way, except on paper. I suppose she'd been trying to explain why she'd done it. But they were her last confused thoughts, and they were all jumbled up, difficult to interpret.' He shut his eyes again and shook his head. 'She couldn't take the deceit any more, she said, knowing that she no longer loved me.

Even if I had been the love of her life. The affair had somehow destroyed all her feelings.' He paused. 'As if it was me who'd had the affair.' He opened his eyes to gaze off into the darkness. 'That's what everyone thought. Including Addie.' He turned his gaze towards Sita. 'Blamed me for cheating on her mother. Driving her to suicide.'

'But there *was* no affair?'

'There was. Only, it wasn't me who had it.' He raised his glass to empty it and found that he already had. He leaned forward to grasp the bottle by the neck and refill the glass before raising it, trembling, to his lips. But the whisky seemed to have lost its malted flavour now. It tasted harsh and burned his mouth. 'Though it didn't look like that at the time. I was partnered with a female detective in those days. Jenny. We were colleagues, mates, but that was all. Jenny came to the funeral with me for moral support, and Addie thought she was my lover. How dare I bring my girlfriend to her mother's funeral!'

He could still feel the sting of her slap, delivered with all the power of pure loathing when everyone had left the house after the wake. Words hurled at him in a fury, barely heard in the moment, and lost now in time. But the shrill tone of anger and accusation still lived with him in every moment of every day. As it would, he knew, till he died.

'She packed all her stuff in a case and left that night to stay with a friend. I haven't seen her or spoken to her since.'

Sita reached through the candlelight in the dark to place a hand over his and gently squeezed it.

He was overwhelmingly touched, feeling his eyes fill, and fought to prevent the tears from spilling. Big, macho Scottish men didn't show their emotions, after all. He raised his glass to his mouth and emptied it in a single draught. And the sound of a glass smashing broke the soft, simmering silence of the hotel.

They were both startled by it. Sita half turned towards the barroom door. 'What was that?'

Brodie blinked away his emotion. 'Must have been Brannan. I'll go and see.' He was almost glad of the excuse to break the moment. He lifted a candle from the table and carried it to the door.

Shadows moved around the walls of the hall as he crossed it to the open door of the dining room. Empty tables stood in rows, draped with white cloth, chairs tipped up, a forest of legs at angles disappearing into darkness. It felt much colder in here, draughty, and the flame of his candle danced dangerously close to extinction. He saw shards of glass on the floor catch its flickering light. Freshly knocked from a table of wine glasses by someone no longer in evidence.

'Hello?' His voice sounded dully in the dark. 'Brannan?' No response.

An icy gust blew out the candle, plunging him into total darkness. He groped for a tabletop to lay it down and searched through the pockets of his open parka for the headlight he had stuffed into one of them earlier. His fingers found the elastic headband and he pulled it out. A loud bang somewhere on the other side of the dining room startled him. He fumbled for the switch on his torch, and bright white light pierced the gloom. He slipped the elastic over his head to free both hands and turned his head to rake torchlight across the dining room. One half of a pair of French windows opening on to an outside terrace lay open, swinging in the wind. As he hurried towards it, Brodie saw wet footprints on the wooden floor. They came fresh from the open door, and returned to it more faintly. Someone had come in from the outside and beat a hasty retreat when Brodie entered with the candle.

Brodie followed the fresh prints from the open door, back across the dining room and into the hall, where they vanished in the carpet. Had someone been eavesdropping on him and Sita in the bar? If so, why? Retreating into the dining room, the intruder had knocked over a glass, smashing it on the floor.

Brodie crunched his way through the broken glass now, heading back to the open door, and stepping out into the snow that lay ten centimetres thick on the wooden terrace. There, the footprints that came and went were crisply imprinted in the fresh fall, and he followed them down the steps and on to the driveway, zipping up his jacket. Snowflakes fell through the beam of his torch as he followed the footsteps through the darkness towards the trees and the football field beyond.

He could feel his heart pounding distantly beneath fleece

and waterproof layers, cold wet snow settling on his bristled head. Up ahead, he saw a shadow darting between the trees. He shouted, 'Stop!' but only succeeded in sending the intruder off at a run. Brodie ran several metres himself into the trees, but quickly realised he would never catch their eavesdropper. There had been far too much whisky consumed. He stopped, breathing heavily for several moments, before turning reluctantly back to the hotel.

Sita turned in her seat as he came into the bar, surprised to see the snow on his jacket. He stamped his feet and shook it off in front of the fire. She said, 'Who was it?'

'No idea. But someone was out there in the hall listening to us talking in here. I don't know how much they could hear, or why they would want to, but they ran off through the snow when I went after them with my torch.'

She stood up, a little unsteadily. 'How did they get in?'

'Through French windows in the dining room.'

'Broke in, you mean?'

Brodie shook his head. 'There didn't appear to be any damage. It couldn't have been locked.' He pursed his lips thoughtfully. 'But we'd better lock ourselves into our rooms tonight. Don't want to offer open invites to any unwanted guests.'

She lifted her bag and crossed to the fire. 'You think we're in danger?'

He shook his head. 'No. I mean, why would we be?'

She shivered, in spite of standing in front of the flames. 'I don't like this place,' she said. 'I've spent half my life with corpses. But the thought of that dead man folded into the cake cabinet in the kitchen gives me the willies.'

Brodie lay on his bed in the dark, fully dressed. He didn't think he would sleep much tonight, and every time he closed his eyes, the room seemed to spin around him. So he stared, unseeing, at the ceiling.

He had never, in all the years since, told anyone about the events of that night when he came home to find Mel dead in the bath. Not even Tiny. He had locked them away tight in a dark place that only he visited. Scared to let the memories escape into the light where, somehow, he felt they would only do him even more harm. He knew exactly why he had not wanted to confront Addie with the truth at the time. She wouldn't have believed him. Wouldn't have wanted to hear it. The man who had betrayed the trust of his wife and daughter just trying to make excuses.

And in bottling it up, he had only made it worse, burying it and damaging himself in the process. Until they had passed the tipping point, he and Addie. That crossroads beyond which there was no return. A time when healing might still have been possible, if only they had made the effort. It wasn't until now, with his own death imminent, that he had been moved, finally, to drag all the skeletons from his closet and lay them out to be judged. Whatever that judgement might be.

He thought of Sita, lying on her own in the next room, cold probably, and a little scared, guarding her private grief behind a bold façade that she had let slip tonight. Unintentionally. To a stranger. And maybe that was easier.

Harder, he thought, to face someone you love with the truth that you've been hiding from them for years.

CHAPTER ELEVEN

Brodie awoke to daylight and a hangover, still fully dressed and surprised to find that he had slept at all. He had not even thought to draw the curtains the night before. Now the reflected light of a white world beyond the glass washed across the ceiling, and he blinked with the pain of it, his head still fuzzy from the whisky.

Slowly, he swung his legs to the floor and stood up, stretching all the stiffness out of his limbs. Outside, sunlight touched the tops of snow-capped peaks as far down the loch as he could see. Garbh Bheinn, Mam na Gualainn, and others. The valley itself languished in the permanent shadow cast at this time of year by the mountains that ringed it, and Brodie saw wisps of mist curl gently upwards from the coruscating surface of the loch.

He recalled Brannan talking of lake-effect snow the night before, and wondered just how much the waters of the loch were warmed by the process of cooling a nuclear reactor. He doubted that it would feel particularly warm if he were to plunge himself naked into it.

An unbroken blue sky lay mirrored in the water. As did the

mountains themselves, reflections shimmering in the gentle breeze that breathed through the fjord and ruffled its surface. It seemed very still out there. The only sign of life was the occasional thread of blue smoke rising from the odd village chimney. There would be few folk left burning wood these days, he thought. Most had converted to geothermal or air source heat pumps. But wood burned from managed forests was thought to equal carbon-neutral. So . . .

He tried the light above the bathroom sink. Nothing. Still no electricity. He slunged his face in cold water and cleaned his teeth with a few perfunctory strokes of his brush, then realised he had forgotten to remove the earbuds of his iCom. Without power, there would be no signal, but he decided to leave them in anyway. He regarded the day's silvered growth on his face and decided, too, not to shave. He would get done what needed done today. The stuff said that needed said. And, power cuts permitting, he would be gone by tonight.

He knocked softly on Sita's door, and when there was no reply, tried the handle. It wasn't locked and he pushed the door open. Like him, she had not slept in the bed, but on it. An impression of her body in the duvet was clearly visible, the shape of her head pressed into the pillow.

He went downstairs and heard voices coming from the dining room. Sita and a young man were sitting at a table set for two. She turned as he came in, her eyes clear and bright, and showing no signs of last night's session with the bottle of Balvenie DoubleWood. 'Oh, I thought for a minute you might be Mr Brannan,' she said. 'He laid out breakfast, such as it is. Some cold meat and a few slices of cheese. But there's no sign of him.'

The young man rose quickly to his feet and Brodie saw that he was in uniform beneath his reflective waterproof jacket. His peaked, chequered cap lay on the table, and he seemed uncertain for a moment as to whether or not he should put it on.

'Constable Robert Sinclair, sir,' he said, extending a hand.

They shook, and Brodie saw that he was a handsome young man. Blue eyes in a fresh, clean-shaven face. A fine, well-defined jawline and an easy smile. Built, too. A good two to three inches taller than Brodie. So this was the man his girl had married. He cast critical eyes over him and said, 'I'm told you're known by everyone as Robbie.'

Robbie seemed momentarily discomposed, a flush of embarrassment on his cheek. 'That's what folk call me, yes, sir. We're very informal here.'

'Good. Most folk call me Cammie.'

'Yes, sir,' Robbie said, without a moment's hesitation. And it was clear that *sir* was the only form of address he would be likely to use. He waved a hand towards the table. 'I brought a flask of hot coffee. We've got an ancient wee camping gas stove at home, and a little gas left in the bottle. I knew they were all electric up here and you would probably be wanting something hot to drink.'

We, Brodie thought, was Robbie and Addie. But all he said was, 'That's very thoughtful, Robbie, thank you.' He pulled up a chair and sat down to pour himself a cup from the flask, and Robbie took that as a signal he could resume his seat.

Sita looked at Brodie. 'Are you not going to eat anything?'

He glanced at the cold meat and the slices of processed cheese curling around the edges. 'Not hungry,' he said.

She grinned. 'Don't blame you.'

Robbie said, 'We've set up one of the surgeries down at the health centre for the PM. Still no power, though.'

Brodie looked at Sita. 'Do you need power for the autopsy?'

'Just a healthy dose of daylight and the power of my elbow,' she said.

'Good. The sooner we get this underway, the better.' Brodie drained his cup and stood up. 'You got a body bag?'

'With my kit.'

'We'd better go get it, then, and move the body to the health centre.' He turned to Robbie. 'We got transport?'

'We've got my SUV, sir.'

'Let's do it, then.'

They drove to the football field in Robbie's SUV. The eVTOL stood mid-pitch where they had left it, heaped with a covering of snow. At the gate, Brodie said suddenly, 'Stop!' Robbie brought the vehicle to a slithering halt.

'What is it?' Sita was alarmed.

'Footprints. Someone's been having a good look at Eve. Wait here.' He jumped down into the snow. A single set of footprints approached the gate from among the trees and tracked off in a determinedly straight line towards the e-chopper. Brodie followed them and saw that whoever had come to take a look at the flying machine had circled it a couple of times, stopping at the doors on each side, perhaps trying to get in. Then they tracked away again towards the far side of the field, and he saw there was a pedestrian gate leading out to a path that headed down towards the river.

He turned and waved to Robbie, and the SUV approached slowly across the pitch. When it reached the eVTOL, the other two jumped out into the snow.

'Was it our intruder from last night, do you think?' Sita said.

Robbie frowned. 'Intruder? What intruder?'

'We had an unannounced visitor at the hotel last night,' Brodie said. 'Came in through the dining room. We were in the bar, and I think he probably stood in the hall listening to us. But he broke a glass on the way out, and that's what alerted us. I followed his footprints as far as the trees but lost him there.'

Robbie was still frowning. 'I don't understand. Why would anyone want to listen to you talking?'

'Good question,' Brodie said. He turned to Sita. 'And in answer to yours, yes, I think it probably was our intruder, coming to give Eve the once-over.'

Sita said, 'Should we get her charging?'

Brodie smiled. 'No power, Sita, remember?'

She tutted and raised her eyes to the heavens. 'Too much bloody whisky last night.' And there was, perhaps, just a moment between them when each remembered the things that the other had confided in vino veritas. 'Let's get my kit.'

They returned to the hotel, taking a black body bag from Sita's Storm case into the kitchen. There still wasn't any sign of Brannan. Brodie called up the stairs but got no reply. Private quarters at the back of the hotel were not locked, but Brannan was not there either.

Robbie said, 'There were tyre tracks on the drive when I arrived. Must have been Brannan's four-by-four. Maybe he's gone into the village for provisions.'

Brodie shrugged. 'Then we'll just have to do what we have to do without his permission.'

They wheeled the cake cabinet from the anteroom into the kitchen and laid out the open body bag on the stainless steel island beneath all the pots and pans and cooking utensils. Brodie lifted the glass lid of the cabinet and glanced at Robbie. 'You okay to do this?'

Robbie nodded, and between them they lifted the dead weight from the cold cabinet to lay along the length of the body bag. The colour of the corpse had changed, even since last night, when it had looked pink and almost fresh. The cake chiller had kept the body frozen to an extent, but in just twelve hours without power it had begun to decompose, skin colour morphing from red-grey to grey-green.

Sita said, 'For some reason, bodies that have been frozen, then thawed, decompose faster than if they'd never been frozen at all.'

'If he'd never been frozen at all, there wouldn't have been much of him left after three months,' Brodie said. He had seen many dead people over the years, but the tiny smile of serenity on Mel's face was the memory that obliterated all the rest. As if she had somehow found peace in death. Conversely, the look on Younger's face suggested fear, or pain, in the moment of dying. The open eyes, the gaping mouth. The skin of his face was broken and contused. If he had been wearing gloves on the climb, they were nowhere in evidence, and the skin of his hands was marbling, as if the blood were leaking from every vein and spreading out beneath the epidermis.

'It was a helluva job getting him into the cabinet when we brought him down off the mountain,' Robbie said. 'I thought it was rigor mortis, but I guess he was just frozen.'

Sita said, 'Rigor only lasts for around three days. You're right, he'd have been frozen solid, stiff as a board, entombed like that in the ice for three months.' She zipped up the body bag and Younger vanished into his now accustomed darkness.

Brodie turned to Robbie. 'You were with the group that brought him down?'

'Yeah, I'm a member of the mountain rescue team. There were a dozen of us went up to get him. Had to chip him out of the ice with our axes. Wasn't easy, lying on your back hacking away at ice just above your head in limited space, being careful not to damage a corpse. Worse, because there's a dead guy staring down at you the whole time. We took it in turns. Then strapped him to a litter and lowered him on ropes, little by little, till it was possible to carry him.'

Brodie nodded. He could imagine just how difficult, and stressful, that must have been. It would be easier getting him into the back of the SUV. 'Let's get him down to the surgery.'

The Kinlochleven Medical Practice stood in a jumble of buildings in Kearan Road, on the far side of the street from the police station, and a prayer away from St Paul's Episcopal Church at the end of the road. The original building had been expanded several times over the last fifty years.

The streets around it were empty, only a handful of tyre tracks in evidence, but Brodie was aware of curtains twitching, and eyes on them as they carried the body bag into the room that had been prepared for them at the back of the building. There was a strange still in the air, thick snow all around absorbing every sound, smothering it in tenebrous silence. A sense here of being shut out from the rest of the world, long mountain shadows casting their gloom on the water, while revealing tantalising glimpses of the world beyond in the ring of sunlight that illuminated the peaks and set them sharp and clear against the blue of the sky.

Brodie was sweating by the time they laid the body on the examination table and unzipped the body bag.

'Just leave him in the bag,' Sita said. 'It'll contain the fluids. Don't want to make a mess on the floor if we can help it.'

She had opened up her Storm case on a table pushed against

the far wall, and was slipping into green scrubs. She donned a heavy apron and pushed her dark, wiry hair into what looked for all the world like a plastic shower cap.

Brodie glanced into her case and saw scalpels, a 35-centimetres chef's knife, forceps, scissors, a ladle, needles, syringes, and a selection of Vacutainers and sealable plastic bags. There were jars of formalin, and paper and plastic evidence bags. Twine and needle. For sewing up the body afterwards, Brodie assumed. He was not looking forward to the flight back to Glasgow, sharing the tiny cabin of the eVTOL with a decaying corpse.

'What else do you have in there?' he said.

'Oh, a veritable Aladdin's cave of goodies. A camera.' She lifted it out. 'You'll be handling that.' And she took out a torch. 'Could have done with this last night. I'll use it to light whatever we need to photograph.' She thrust it at Brodie. 'Also have a handheld X-ray machine. It can do arms and legs and heads. Not big enough for the torso, though.' She pulled on plastic shoe covers and snapped her hands into latex gloves.

'You come well equipped. No wonder this thing was heavy.'

'Got to think of everything.' She grinned as she lifted out a surgical handsaw. 'In the absence of electricity, we're going to have to open up the skull the old-fashioned way.' She turned towards Robbie, who was standing by the door looking pale. 'I'm going to cut him out of his clothes first, and you two can lay them out on the table over there. There's a roll of paper in my case that you can spread out on it.'

'Me?' Robbie seemed shocked.

'You are staying for the PM, aren't you?'

'Well, I . . . I hadn't really thought . . .'

Brodie said, 'First one, son?'

Robbie's eyes darted self-consciously in his direction and he nodded.

Sita laughed and said, 'Well, it probably won't be your last. Got to start somewhere.' She stopped and thought for a moment. 'Something useful you *can* do. Go home and bring me a plastic bucket for the excess fluids. And a stainless steel bowl if you have such a thing. I need something to put the organs into before I dissect them. Oh, and if you've any gas left in that old stove of yours, you could heat me up some water. I'm going to need to thaw out my hands from time to time.' She turned towards the body. 'This fella's still going to be pretty damned cold inside.'

Brodie caught Robbie's arm as he turned to go. 'I believe it was your wife who found the body,' he said. Robbie nodded. 'I'm going to need to talk to her. And I'm going to need somebody to take me up to the snow patch where it was found.'

'Oh, Addie'll do that. She's scheduled to go up again anyway for a routine maintenance check on the weather station after the storm. I'll speak to her when I go over to the house. She can come across when the post-mortem's finished.'

Brodie nodded, and felt his heart rate rise.

*

Younger's clothes, all laid out now on the table, were torn in places, badly abraded in others. An anorak over a fleece. Ski pants. His leather boots were badly lacerated, the uppers on one of them ripped completely free of the sole. Sita held the torch as Brodie photographed them.

She packed towels around the body, and got Brodie to photograph it as well. She was particularly interested in the face. 'Look at these,' she said, running a latexed finger over irregular-shaped random contusions and abrasions. Most were broad brush-type abrasions, several of them appearing over the prominences of the face, around the eye sockets and high parts of the cheeks. Similar injuries were in evidence, too, around the rest of his body, but less severe where he had been protected by his clothes.

Brodie nodded. 'Injuries from a fall?'

'Looks like it.'

'An accident, then?'

'Not so fast. Look closer.' As Brodie leaned in to examine Younger's face, she said, 'See? There are multiple blunt force injuries, different from the others. Look at the left cheek. There are seven sets of patterned injuries consisting of four short, parallel abraded contusions, about 3.8 centimetres in length and 0.4 millimetres apart. And check out the single faint linear contusion running perpendicular to the groups.'

Brodie could see that the injuries she was describing formed some kind of a pattern. 'What do they mean? How did he get those?'

She looked up and smiled from behind her mask. 'Someone hit him, Mr Brodie. Punched him. Someone wearing a very distinctive pair of gloves. Gloves with some kind of protective reinforcement along the backs of the fingers, notched with four horizontal niches at each knuckle to allow the fingers to flex, and a raised ledge running along the length of each finger.' She moved her fingertips to Younger's forehead. 'Two more here as well. And another along the right jawline.'

'Is that what killed him?'

'I doubt it. Enough to knock him off his feet, though. Cause him to fall, which would be consistent with his other injuries.'

Robbie came in with a basin of steaming hot water. 'This'll be too hot to put your hands in just yet.'

'Put it on the table over there. I won't need it till I cut him open.' She lifted one of Younger's hands and examined it closely, turning it this way and that, then fetched a tiny scalpel and a piece of paper torn from a notebook, before gently scraping residue from beneath the fingernails of the right hand to collect on the paper. 'I think,' she said, 'we'll find that this is skin. Almost certainly harvested from his attacker's face or neck. I'd say our man put up a bit of a fight. He'll have left his mark.' She let the scrapings slide from the paper into a plastic sample bag and sealed it.

Brodie said, 'You'll get DNA from that?'

'We will.'

'How soon?'

'As soon as we get power. The wonders of technology. We

have a very smart little piece of kit these days that can do on-site DNA analysis. And assuming we have power, then we'll also have internet access, and I can run it through the database.'

'And cause of death?'

'You know as well as I do, Detective Inspector, that no pathologist worth her salt is going to speculate on that until the autopsy is complete.' She turned towards Robbie. 'Do you have that bucket and stainless steel dish?'

'I'll just dash back across the road and get them.' He hurried to the door and paused there. 'My wife will be over in about an hour, sir, if that suits.'

He nodded. 'Sure.' And he turned away quickly to focus on Sita's scalpel as she made her Y-incision in the body, cutting from each shoulder to the breastbone and then all the way down to the pubis. Although he was losing the hair on his head, Younger had plenty of it on his body, a tangle of wiry fair pubic hair on his chest and belly and back, and the fluids of his autopsy ran freely through it.

It took Sita the best part of three-quarters of an hour to open him up and remove his organs one by one, transferring them to the stainless steel bowl that Robbie had brought to an adjoining table, where she carefully bread-loafed each one. After Robbie returned, he had stood at the far side of the room watching at a distance, white face tinged now with green.

Sita asked the two men to leave the room while she cut around the skull with her handsaw. 'We don't want to be breathing in any particulates, now, do we?' she said, double-layering her own surgical mask and slipping on a pair of goggles.

Brodie and the young constable stood outside for some time, stamping their feet to keep warm. 'Do you want to come across to the station for a coffee?' Robbie asked him eventually.

Brodie shook his head. 'Better stay around in case I'm needed.'

Robbie nodded and they stood in awkward silence for some more minutes.

Then Brodie said, 'How long have you been here?'

'Since I was twenty-three. So about seven years, I guess. I was at Inverness before. Then this posting came up, and I thought, why the hell not? I grew up myself in a village near Fort William. I like the informality of village life.' He shrugged. 'Lifestyle's much more important to me than career. I mean, I guess I might have thought about moving on, climbing the ladder, but then I met Addie.' He allowed himself a fond laugh. 'And, well, here I still am. Got a young kid now, too, so that makes a difference. You got family?'

Brodie couldn't meet his eye. He just nodded, and breathed out slowly. 'Yeah.'

Robbie waited for Brodie to tell him more, but when no further elucidation was forthcoming, they fell into an awkward silence, and Brodie was relieved ten minutes later to hear Sita calling from the surgery. They went back in.

She was peeling off her latex gloves and freeing her hair from its plastic protection, shaking it free to tumble over her shoulders. On the table behind her stood an array of jars and plastic bags with all the samples she would take back with her for laboratory analysis. The body was all sewn up, the skull cap replaced, and Younger looked as if he had just been carried off the set of the latest Frankenstein movie.

Finally she broke the silence she had maintained throughout most of the post-mortem. Ready to pronounce on cause of death. 'Disarticulated vertebrae in the neck,' she said. 'Cut the spinal cord clean through. That would certainly have killed him, even if the multiple fractures of his skull hadn't. Both forearms broken, right tibia. It was quite a fall, I think.'

'As a result of the blows struck by his attacker?' Brodie said.

'Well, we can speculate on that. But all I can say for certain is that he was in a heck of a fight before the fall.' She started to remove her apron, then paused. 'There's some other stuff, though. Weird stuff that I can't quite explain.'

'Weird in what way?'

'It might not even be related.' She thought about it some more. 'There was sloughing off of the gut mucosa. With a fair bit of inflammation. In the lungs, too. I mean, with a big fall like he had, pulmonary contusion would be possible.' She paused to explain. 'Lung bruising. But because he died pretty quickly, there wouldn't have been any accompanying inflammation. I sampled some random areas of the lung for microscopic examination. And there was plenty of haemorrhaging and inflammation, which I really wouldn't have expected to see. It doesn't fit with trauma, or being frozen.' She shrugged and smiled. 'Can't know everything. But I'll get some detailed analysis done on the samples.'

A tentative knock at the door brought colour to Brodie's face, and his heart beat faster.

A young woman's voice called, 'Are you finished in there?'

Robbie turned towards the open body bag. 'Can we . . . ?'

'Of course,' Sita said, and zipped it up to conceal Younger from innocent eyes.

Robbie crossed the room to open the door and Addie stepped in. She seemed hesitant. Her smile was uncertain. She said, 'Hi.'

Addie had barely changed in all the years since Brodie had last set eyes on her. A little older. The faintest evidence of crow's feet at the corners of her eyes. She was carrying a little more weight. But then, she'd had a baby. She still looked fit, though. All that climbing up and down Binnein Mòr, and the other mountains in the Mamores where she had installed her weather stations. Her hair was the same silky chestnut brown falling to fine, square shoulders. Her eyes and mouth were still Mel's. He had always seen more of her mother in her than himself. If she had inherited anything of him, it was his temper.

She looked around the room, nodding acknowledgement to Sita, and then her eyes fell on her father. He saw the momentary confusion in them as she processed disbelief, which morphed to realisation, and then to anger. It didn't take long.

'What the fuck . . . ?' An almost involuntary exclamation under her breath. Then the explosion. 'What the hell are you doing here?'

Robbie was startled. 'Addie!'

But like a terrier following a scent and deaf to its owner's calls, she ignored him, focused entirely on her father. She was shaking her head. 'This can't be a coincidence. You must have known. You planned this, didn't you?'

Brodie was surprised by the calm he heard in his own voice. 'Nobody plans for murder, Addie.'

Robbie cut in, perplexed. 'Wait a minute. You two know each other?'

Addie still wasn't listening, but was deflected by the word *murder*. She flashed a look at Sita. 'Murder? That man I found was murdered?'

Sita was startled by this unexpected turn of events, and nodded mutely.

Addie was stopped momentarily in her tracks. But it didn't last. She freed herself of the thought and turned blazing eyes back on her father. 'Why? Why now, after all these years? What did you think? That I was going to throw my arms around your neck, and say, *Daddy, everything's forgiven*?'

Robbie dragged his gaze away from his wife and turned it towards Brodie with incredulity. 'You're her father?'

Brodie was embarrassed. 'I'm sorry,' he said. 'I should have told you.'

But nothing was going to stop Addie. 'Oh, yes, *sorry*! That's you all over, isn't it? Always sorry.'

Robbie stepped in firmly, embarrassment giving way to

anger. 'Addie, stop it!' He took her by the shoulders, but pulled up short of shaking them. 'I don't know what's going on between you two.' He drew a sharp breath. 'Because, let's face it, you've always told me that both your parents were dead.'

She tore her eyes away from Brodie, and a fleeting moment of guilt diluted the anger in them.

Robbie said, 'This is a murder investigation, for Christ's sake. You're a material witness. And like it or not, you're going to have to take your father up the mountain to show him where you found the body. Now, I suggest you get a hold of yourself, go home and get changed for the climb.'

She glared at him with naked hostility. 'Whose side are you on?'

'I'm on the side of the law, Addie.' He made a determined effort to lower the pitch of his voice. 'Now go and get changed.' He let go of her shoulders.

She stood trembling with anger and humiliation. Then turned her eyes beyond her husband to settle again on her father. 'See?' she said. 'All these years I've been happy without you. You're back in my life for two minutes and causing conflict already.'

As she turned to the door, one of the gloves she'd been clutching and twisting in her hands fell to the floor. But she wasn't about to ruin her exit, and ignored it as she stomped off through the snow. Robbie was too embarrassed to notice. He half turned towards Brodie, barely able to meet his eye. 'I'm sorry about that,' he said. 'I'll speak to her.' And he hurried out into the chill of the morning in pursuit of his inexplicably hostile wife.

Brodie stepped to the door and stooped to pick up the glove. Soft, hand-sewn lambskin, turned over at the wrist. It was still warm, and for a moment it felt like holding her hand. He raised it to his face and breathed in her scent deeply before closing the door. Then he turned to find Sita staring at him. Concern was etched deeply in the lines around her mouth, and reflected in the light that diffused the darkness of her eyes. 'Your daughter? Really?' She hesitated. 'Of course, you knew?'

He nodded and she closed her eyes.

'For God's sake, Brodie. I mean, she's right. What on earth did you hope to achieve?'

He hadn't achieved it yet, and he wasn't about to tell her.

'Do they know? In Glasgow, I mean. Your bosses?'

He shook his head. 'No.'

She sighed in frustration. 'They would never have sent you if they had. And you should never have volunteered, if that's what you did. Your daughter found the body. She's a potential suspect.'

'Addie didn't kill anybody.'

'You don't know that. No one knows that.'

'You think she's big enough and strong enough to punch a man the size of Younger off the top of a mountain?'

'No, of course not. But that's not the point.'

'What is?'

'That you should not be involved with this investigation in any way. You have to declare a family interest. They'll send someone else.'

'We have no power, remember. No comms. No way to contact HQ. So I'm just going to have to make the best of it.'

She stared at him for a long time, the slightest shaking of her head. 'Why did you come?'

'There are matters I need to settle before . . .' His voice tailed away. 'Just things I need to settle.'

The slightest cant of her head, the faintest narrowing of her eyes, posed a question that she didn't frame in words. Perhaps suspecting that there would be no answer forthcoming.

Brodie looked at Addie's glove in his hands and said, 'I've heard that sometimes gloves can be a good source of DNA. A tear in the cuticle, a spot of blood dried into the lining.' He looked up. 'Is that right?'

She frowned. 'It's been known.'

He took a step towards her and held out the glove. 'Any chance you could look for a sample in this?'

Now she was incredulous. She took the glove. 'You just told me there's no way you think she's involved in Younger's murder.'

He scoffed. 'Of course she's not.' He crossed the room to where he had draped his parka over the back of a chair, and turned the hood inside out. There were quite a number of hairs trapped in the fleece from a time before his razor cut, when his hair had been longer. He teased some of it free and held it out to her. 'If you find some, maybe you could check it against mine. See if there's a familial match.'

'You think there might not be?'

'I'd just be grateful if you could do that for me.' He paused. 'Can you?'

She took the hair and slipped it into a resealable evidence bag. 'You sure you want to know?'

He pursed his lips, and she saw the sadness in his eyes as he nodded, almost imperceptibly.

CHAPTER TWELVE

Brodie walked back to the hotel, following the tyre tracks on the B863. He crossed the bridge spanning the stream they called Allt Coire na Bà, where it ran in spate down from Grey Mare's Waterfall before joining the River Leven as it debouched into the loch. Across the valley, the windows of the school simmered darkly. Absent of the sound of children's voices. There were no footprints breaking the surface of the freshly fallen snow on the school playing field. No power, so no school. Frustrated schoolkids no doubt sitting at home staring at blank TV screens, unable even to fire up games on their PlayStation Fifteens. No evidence, either, of them playing outside. Perhaps they had forgotten how.

Robbie had told him he would bring Sita and the body, and all her kit and samples, back to the hotel once she had cleaned up. Brodie wanted to get up on to the mountain before the light began to fade.

Brannan's four-by-four was nowhere in evidence when Brodie reached the International. He pushed open the main door, kicked the snow off his boots, and walked into the hallway. It was silent as the grave in there. Gloomy without any direct sunlight spilling through windows. He called out, but there was no response. He was hungry, but there was no time to go foraging for food. Instead, he climbed the stairs and went into his room to prepare for the mountain.

He pulled on elasticated stretch pants over his long johns, and a microfleece top over a synthetic base layer. The weather was dry, with no imminent risk of further snow, so he would wear his down-filled North Face parka on top of that.

He sat on the bed to pull on a pair of stiff-soled B2-rated mountaineering boots, and attach the snow gaiters that would keep his lower legs dry. His articulated C2 crampons lay on the duvet. He would put them in his pack and attach them to his boots when they emerged from the woods to begin the climb up through the snow.

His gloves, which extended to cover his forearms, were a halfway house between a glove and a mitten, with separate sheaths for thumb and forefinger. He stuffed them in his pack, and before pulling on his woollen hat to cover his ears, caught sight of himself in the mirror above the sink. Unshaven, complexion like putty, salt-and-pepper silver stubbled hair. The face with which he had greeted his daughter for the first time in ten years. And he thought how old he looked, and weary. And in that brief encounter had felt how much she hated him still.

Robbie had promised that Addie would meet him at the Grey Mare's car park. He half expected she wouldn't be there, and half hoped he might be right. He was, he realised, dreading the climb with her. He had no idea what he was going to say. Had rehearsed nothing. Taken the decision to come on the spur of the moment, and like a marriage made in haste, was regretting it at leisure. But he also had a job to do. A man had been murdered. Outside help was not an option, since he had no way of contacting Glasgow. So he was on his own. In more ways than one.

He was in the downstairs hall when the power came back on. Lights flickered to life in the dining room, and he heard the refrigeration units in the kitchen kick in. He checked the time. It was approaching midday. He swithered briefly about whether or not to check in with Glasgow and report Sita's findings. Instinctively, he touched his breast pocket to check that his iCom glasses in their protective case were still there. He decided against making the call. It would only delay him. And complicate things. He needed the time with Addie that the climb would give him, and would call when he got back.

He left the hotel and made his way through the trees to the football field. Now that the power was back on, he could get the eVTOL charging for the return journey. As he walked through the gate on to the pitch, he stopped. There were more tracks now than previously. Robbie's tyre tracks had obliterated the initial single set of footprints leading out to the e-chopper that they had spotted earlier. He could see where the three of them had got out of the vehicle to recover Sita's Storm case. And the original set of prints that had circled Eve before heading off to the smaller gate on the far side of the field. Now a second set of prints came from that same gate and circled the chopper before disappearing among the tyre tracks towards the pavilion outside the main gate. Perfectly possible, of course, that it was just some curious local, though Brodie reflected he had seen precious few folk out and about on this morning after the storm. He circled the eVTOL himself to check for damage, or any sign of forced entry. But there was nothing.

He sighed and opened the hatch to access the charging cable, and tracked off with it across the field towards the pavilion. There he found the charging hub and plugged it in. It seemed like an archaic process, but he figured it was probably just as efficient as wireless charging. Lights on the reader attached to the plug unit flashed green, which satisfied him that Eve was taking a charge. And piercing unbroken snow with the point of his walker's ice axe, he set off with nervous trepidation for the rendezvous with his daughter.

Addie was waiting impatiently in the car park, stamping her feet to keep warm. She was wearing blue ski pants and a bright yellow parka and woolly hat, hair spilling out from beneath it, almost red in the early afternoon light. Her daypack looked like it wouldn't have much more in it than her crampons, and maybe a flask of something hot. Her ice axe dangled by a strap from her wrist, and from the black look on her face as she saw him coming, Brodie thought that she was probably fighting the urge to bury it in his chest.

But she was accompanied by a group of men in climbing gear clustering around a minibus, laughing and stamping their feet also, breath billowing around their heads in the icy air. Too many witnesses for murder, he thought wryly. They turned to look with interest as Brodie approached. He nodded. 'Gentlemen.' They murmured uncertain greetings in response. He turned to Addie. 'We're all going up?'

'No,' she said quickly. 'I get an allowance from the SMO to pay locals to check on the other weather stations along the Mamores when I can't do them all. And we have a very narrow weather window today.'

'So this is your dad, then?' An older man with a leathery, wind-burned face looked at Brodie with curiosity, and Brodie thought how fast word travelled in a small community.

'Yes, Archie, it is.' Addie looked as if each word was leaving a bad taste in her mouth. Then she turned to Brodie. 'Archie leads the mountain rescue team.'

Brodie leaned forward to shake his hand. 'So, Mr . . . ?'

'McKay.'

'You were in the group that brought Mr Younger's body down from the mountain?'

'I was that,' Archie said. 'Most of us here were. Fucking idiot. Took the hard way up when I clearly told him it was not a route for beginners.'

Brodie frowned. 'I understood no one knew where he had gone.'

Archie looked uncomfortable. 'Well, I didn't know that's

where he'd gone the day he went missing. I spoke to him the day before. It was Mike Brannan up at the International who sent him to me for advice. Not that he was inclined to take it.'

'So how do you know which way he went up?'

He coloured a little. 'Well, I don't. But he told me the easy way would take too long. If he went up at all.' He paused. 'It's no wonder he fell.'

Brodie said, 'He didn't fall, Mr McKay. Someone punched his lights out and pushed him off the summit. It's a murder I'm investigating here.'

An almost audible sense of shock rippled through the small group of climbers.

'Who do you think did it?' another of the men asked.

'That's what I hope to find out. But we have a sample of the killer's DNA. So it probably won't be too long before we do.'

Archie said, 'How does that help? I mean, what if the killer's not on a database somewhere.'

Brodie raised an eyebrow. 'You seem to know a lot about it, Mr McKay.'

Archie shrugged. 'I read detective books like everyone else, Mr Brodie.'

'Then you'll know that if he's not on the database, we'll have to DNA test every male in the village.'

One of the younger men said, 'And if someone doesn't agree to that?'

Brodie looked at him. 'And you are?'

'He's my boy,' Archie said defensively.

'And presumably he has a name. And a tongue in his head to answer for himself.'

Archie swallowed his annoyance.

'Tam,' the boy said.

'Well, then, in answer to your question, Tam, he'll be arrested for obstruction.'

Addie intervened for the first time. 'How do you know it wasn't a woman?'

Brodie turned towards his daughter. 'I don't. But Younger wasn't a small man, and his killer gave him a bit of a hiding before he fell into the corrie. It would need to be a pretty powerful woman.'

The group of climbers shuffled uncertainly in the sunshine until Archie said gruffly, 'We'd better be going if we want this done before sunset.' He turned towards Addie, ignoring her father. 'See you later, lass.'

And Brodie had the impression that all of them were glad to escape the moment, clambering into their minibus before heading off to climb their allocated peaks in the Mamores. He watched thoughtfully as the vehicle set off along the B863, but when he turned back towards Addie, found her glaring at him with unconcealed loathing. She'd had time to process his arrival. Time to let her anger build. He braced himself for the storm.

'You selfish fucker! What on earth did you think you could achieve by coming here like this? I'm sure they didn't send you. Not if they'd known. They wouldn't. You must have volunteered.'

Brodie tried to maintain a semblance of calm. 'Addie . . .'

But she cut him off. 'Don't use my fucking name. Don't you dare. Did you think for one minute how I would feel? Did you? Why I haven't spoken to you in all this time? Of course not. Because the only person you ever think about is you. And you never were good on consequences, were you? Mum would still have been here otherwise.' She paused to draw a quivering breath. 'You make me sick!'

And she turned to stride off up the path and into the woods. At a pace he knew he was going to have trouble matching.

They passed signposts for Spean Bridge and Corrour station, and she turned left where the path split at a T-junction, then forked right on to a rougher path up steps. The snow was sparser here beneath the trees, and slippy underfoot.

The path climbed steeply through the woods and she kept up an unrelenting pace, not looking back once. She forked left across a burn and pushed on up through more deciduous woodland. She had an easy, loping stride, and he saw her breath condensing in the sunlight ahead of him.

He paused to catch his own breath, hearing it rasp in his chest, and looked back the way they had come. Already they had achieved a considerable elevation, and a spectacular view of the village and the loch lay spread out below them, blue water zigzagging off between snow-covered peaks towards a sea lost somewhere in the misted distance.

Addie's voice rang out in the cold from up ahead. 'What's wrong, old man? Can't hack it any more?'

He turned to see her glaring down at him through the trees and he sighed, and started off again, following in her footsteps. She stood watching him for a few moments.

'You need to keep up if we're going to beat the light,' she said. 'When the sun starts to go down, it goes down fast.'

Again, she didn't wait for him, turning to push on through the trees along some old stalker's path. It wasn't long before they left the woods behind, and undulating moorland opened up before them. Brodie stopped again, this time to attach his crampons for better grip in the snow, and when he looked for her up ahead saw that she had done the same, her eyes hidden now behind dark glasses. He took off his gloves to fumble in his pocket for the case that held his iCom glasses, and slipped them on. 'iCom, shade my lenses,' he said, and felt foolish, as if talking to himself. But the lenses immediately cut the glare of the snow. 'Darker,' he said, and now he could see without screwing up his eyes. He pulled on his gloves and set off after her once more.

They crossed a stream, Addie still a good fifty yards ahead of him, then turned up the far bank before climbing around the southern flank of Sgùrr Èilde Beag. Away to their right, sunlight reflected in diamond clusters on the dark waters of Loch Èilde Mòr. It wasn't long before he realised that he was starting to gain on her, as if the pace that she had set to defeat him was too much for her. And he was getting his second wind.

They were cutting diagonally across the steep slope of the hill, the odd copse of fir trees breaking the monotony of the snowy wastes. And finally he was at her side, matching her stride for stride. He heard her laboured breathing, though whether it was from exertion or anger he couldn't tell.

He swung his gaze around what was an almost featureless landscape and said, 'How do you know where we're going?'

She took a moment or two to respond. 'I've walked this so many times in all seasons, following the same stalker's trails. I know every feature of this land by heart.' Then, as if annoyed with herself for even speaking to him, she stopped abruptly, turning in anger. 'Why *are* you here? Really? And don't tell me it's your fucking job.' He almost smiled at her propensity for cursing, just like his own. 'I mean, what can you possibly hope to achieve?'

'We need to talk, Addie.'

'No, we don't! I haven't needed to talk to you for ten years, and I've no intention of starting now.'

'Please, just hear me out.'

'No! And don't you dare tell me that somehow I owe it to you. I owe you nothing. Betrayal doesn't deserve forgiveness. Because that's what you did, you know. Betrayed us. Both of us. With that . . .' She searched for a word that would give full force to her contempt and loathing, but came up short. 'How could you bring her to the funeral, how could you?'

'Addie, it wasn't like that. She was my partner at work. She was only there for moral support.'

'And you're going to tell me that you didn't have a relationship with her?'

He let his head fall. 'Only afterwards. And that was a mistake.'

Addie was scathing. 'Oh, so you didn't live happily ever after, then?'

He looked up to meet her eyes, but they were hidden behind the lenses of her sunglasses. 'No, we didn't. Jenny wanted it, but I couldn't. She moved in, but it didn't last six months. When she left, she said there was no way she could compete with a dead woman.'

For a moment Addie was at a loss for words.

'I could never be free of Mel. Or my guilt.'

And his daughter's defiance returned. She removed her sunglasses to glare at him. 'So if you weren't having an affair, what did you have to feel guilty about? I mean, what are you saying? That my mother's last words were a lie?'

It was almost painful to look at her. As if it was Mel standing there on the side of a mountain railing at him. Her eyes, her mouth. His temper. He wanted so much to take her in his arms and tell her he was sorry. But sorry, he knew, wouldn't cut it with Addie, so he kept it to himself. And she turned abruptly, replacing her sunglasses and pressing on up the slope. She was the one who had the second wind now.

As the slope grew steeper, they began to zigzag until they reached a ridge just short of the first minor peak of Sgùrr Èilde Beag. Brodie stopped to catch his breath and take in the view. Already it felt like they were approaching the roof of the world.

The land around them rose and fell in snow-covered splendour as far as the eye could see. Directly across from them, Sgùrr Èilde Mòr rose to its majestic summit, and sunlight glanced off the deep blue of the loch far below.

Addie paused, too. Though she must have seen them many times, he saw the wonder in her gaze as she cast it across the mountaintops. 'Always takes my breath away,' she said, forgetting for a moment that she was talking to him. And then, self-consciously, she turned to press on along the line of the ridge.

Steep, snow-covered slopes dropped away left and right now, the ridge itself still rising and running out to Binnein Mòr. In the distance stood Ben Nevis, and they saw the shadows that cast themselves in deep, dark blue to the east and north of the Grey Corries.

The land dipped away slightly to a bealach, a mountain pass, before rising again along the narrowest of ridges that curved around to the peak, the gradients on either side of them falling almost sheer away. Brodie was glad of the crampons that bit into the crusted surface of the snow, and he allowed himself to tip a little of his weight on to his walker's ice axe to keep his balance and fight off the temptations of vertigo.

He loved the mountains for their sense of solitude, and the context they gave to the problems of his life, which seemed so much smaller up here than when he returned to the turmoil of life below. But he was unaccustomed to having company, and for the first time, it felt like he had brought those problems with him, and somehow the presence of his daughter was magnifying that.

A strong breeze blew in their faces now. The icy breath of the Arctic. His eyes watered and his face grew stiff from the cold. He was glad when finally they reached the weather station and he could stop to regain some equilibrium in his breathing and let his legs recover a little. Right now they felt like jelly, and he wondered how they were going to carry him back down the nearly four thousand feet of this highest mountain in the Mamores range.

He was surprised how small the installation was – a flimsy tripod bolted to the rock, sprouting sensors and solar panels and aerials. He watched her remove her sunglasses, then kneel down to clear away the snow and check its components carefully. 'How in God's name does that survive up here?'

'It's based on the one they built for Everest. So Binnein Mòr's a doddle.'

'And you were up here checking on it the day you found the body?'

'That's right.'

'And there's another four or five of them that you check on across the range? The ones your helpers are doing today?'

She kept her focus on what she was doing. 'Six in total. Did you read that in your briefing notes?'

He squatted down on his haunches. 'You'd be surprised how much I know about you.'

Her fingers froze for a moment on the lid of the box she had opened to check on the battery, and she cast him a sideways glance. 'Oh?' She closed the lid. 'Like what?'

'I know you quit Glasgow University after six months. A sudden change of mind. Went to Edinburgh instead. Did an honours degree in meteorology. Then got a job at the Scottish Met Office.'

She turned to look at him directly now. If anything, it seemed that the hostility in her gaze had intensified. He was almost discomposed by it, but pressed on.

'You came up here, leading a team to install these weather stations along the Mamores. Which is when you met Robbie. The local cop. And fell for him.' And he added wryly, 'Not, I'm quite sure, as the result of any kind of father fixation.'

She stared at him in silence for what felt like a very long time. 'So you've been spying on me. Like some kind of stalker.'

'Is it wrong for a father to take an interest in his own daughter? Especially when she's never going to tell him anything about her life herself? A daughter who never invited him to her wedding. Didn't even tell him she was getting married.'

'But you knew anyway.'

He nodded, and returned the intensity of her gaze. 'What I didn't know was that you'd given up full-time work to have a kid.' And he saw anger flare in her eyes.

'So that's why you're here?'

He shook his head. 'No.'

But she didn't want to hear it and was back on her feet in an instant. 'You have no claim on my son. I don't want you anywhere near him. He doesn't even know he *has* a maternal grandfather.'

The hurt must have shown on his face, for there was a fleeting moment of regret in hers. To cover it, she swung her pack off her back and delved inside for a flask and two tin cups. She pushed the cups into the snow to stop them from blowing away and poured hot, milky coffee into each. Without a word, she held one out towards him.

He took it gratefully and felt the hot liquid warming his insides. He stood up to stop his legs from cramping and nodded towards another installation perhaps fifty metres further along the ridge. 'Another one of yours?'

She turned to look. It was smaller than her weather station. Anchored in a similar way, with some kind of sensor on top of a tall pole to which there were two large solar panels attached above a battery box and reader. 'No. That's a GDN field installation.'

He shrugged, none the wiser. 'Which is what?'

'GDN. Gamma Detection Network. There's a ring of about sixteen of these things set up in a ten- to fifteen-mile radius of the power plant at Ballachulish A. They monitor radiation levels. Nothing to do with me. Someone's probably already been up from the plant to check on it after the storm.' She drained her coffee and stuffed the cup back in her pack, and held out her hand for his. He finished the last of his coffee and passed it to her. She said, 'It's time we went down into the corrie.'

*

They stood on the lip of the drop into the Corrie of the Two Lochans and felt how the wind had picked up. Brodie planted his legs well apart to keep his balance.

Addie braced herself too, but shook her head as she gazed down into the deep hollowed cavity on the north side of the mountain. 'There's been so much snow in the last few days,' she said. 'I can't even see the ice tunnel now. I don't know if I'll be able to find it again.'

There were two ridges flanking the corrie, and they took the one on the west side. It was steep, very nearly sheer in places, and their descent was slow and careful, leaning towards the perpendicular using their crampons for grip and their ice axes for balance.

As they descended into the shadow on the north side of the mountain, they felt the temperature drop, and Brodie paused, bracing himself against the angle, to remove his iCom glasses and slip them back into their protective case. When he looked up again, he saw Addie watching him.

She turned away quickly and they slithered down the side of the ridge and into the corrie itself, to traverse the snow that had gathered thickly in the hollow. Breathlessly, Addie said, 'It was somewhere over here. Right in the deepest part.' She stopped to scan the contour of the slope. 'You know, thirty years ago, snow hunters used to scour the mountains for snow patches that survived throughout the year.'

'Why would they do that?'

She shrugged. 'Who knows? To note and monitor them for

posterity, I guess. The thing is, there were precious few of them around back then. And as global temperatures rose, they vanished altogether. Gone by the late spring. Now there's hundreds of them all over the mountains, lying in deep corries just like this one all year round.'

Brodie squatted in the snow, using his ice axe to keep his balance. Her change towards him was small, and subtle, but hadn't gone unnoticed. At least she was talking to him. 'That's something I've never really understood,' he said. 'How it got cold here and hotter nearly everywhere else.'

Her look was scathing. 'Probably because, like everyone else, you just weren't paying attention.' He felt the sting of her rebuke. But she wasn't finished. 'Bet you didn't even care to know. Certainly didn't care enough to do anything about it.'

'Maybe you'd like to explain it to me, then. Since you're the one with the degree.'

She detected and reacted to his sarcasm. 'It's perfectly simple. Simple enough even for you to understand. You've heard of the Gulf Stream, I suppose.'

'Of course.'

'Yeah, well, it pretty much doesn't exist any more. It brought warm water from the Gulf of Mexico north-east across the Atlantic. The whole of western Europe was warmer as a result. Particularly Scotland. I mean, if you look at other countries on the same latitude as Scotland, you'd see that snow and ice are the norm. Basically we line up with the whole of the Alaskan panhandle.'

She exhaled through pursed lips, and Brodie saw that there was an anger simmering deep inside her.

'When the Greenland ice sheet started melting, all that freezing meltwater plunged south and basically stopped the Gulf Stream in its tracks.' She paused. 'It got colder. And then there's the jet stream. I suppose you know what that is, too?'

'I've heard of it.'

'A stream of air circling the northern hemisphere. Caused by warm air rising from the equator, meeting cold air dropping from the Arctic. It used to be that if the jet stream sat higher than usual, we would have a good summer. Lower, and it would be crap. But when global temperatures started rising, the air from the equator got hotter and disrupted the flow of it. Deforming it into peaks and troughs. The peaks drew up even hotter air, and the troughs pulled cold air down from the Arctic, from a circulation up there called the polar vortex. Put everything together, and suddenly Scotland's got the climate of northern Norway, and the equator's so fucking hot, no one can live there any more.'

She took a deep breath as if to try and calm the passion that was the cause of her agitation. And she turned further recrimination towards her father.

'That's what happens when you don't fucking listen. That's the legacy your generation left mine.'

Brodie stood up as he felt anger spike through him. The temper that he had passed on to his daughter. 'Oh, I was listening. Like everyone else. It was practically all you ever fucking heard about. Climate change. Global warming. How we all had to do our bit. And a lot of us did. But the big boys didn't, did they? China, India, Russia, America. The economic imperative or something, they called it. The need to keep on sucking fossil fuels from the ground and burning the fucking stuff, because too many people were making too much money doing just that.' He waved his ice axe towards the heavens. 'And what could ordinary folk like me or you do about it? Fuck all. It's like when they tell us we're going to war. Or they're going to spent billions on nuclear weapons. Or refuse entry to starving immigrants. Whether we agree with any of it or not.'

'You could have taken to the streets.'

He breathed his scorn into the wind. 'Oh, yeah, that works. Disrupt the flow of daily life and people get pissed off with you. Protest in sufficient numbers and the authorities send in the riot police. You get one chance to change things, Addie. Once every four years. You put the other lot in, and it turns out they're just the same.' He rammed the point of his ice axe into the snow. 'In the end, that's why I stopped listening. Stopped caring. And it doesn't matter what generation you belong to, nothing changes. It's the same people abusing the same power, and making the same money.'

He found that he was breathing hard now, shocked by a passion he didn't know he possessed. She was staring at him. But it wasn't hate he saw there. She was startled. He grew suddenly self-conscious and tried a smile that didn't quite work.

'Don't know where that came from.'

She stood staring at him for a moment longer, then turned away suddenly. 'I'll see if I can find that ice tunnel.'

It was a depression at the bottom end of it that gave away its location. Buried under fresh snowfall, it still presented a slightly raised profile at the upper end, which fell away sharply where the entrance to the tunnel had caused the wind to eddy and scoop out a hollow in the snow.

'This has got to be it.' Addie started dragging snow away with her gloved hands, and Brodie crouched down to join her, scraping the top layer away until the snow above them slid off the mound, and suddenly the entrance to the ice tunnel was revealed.

Brodie fished out the headlight from his pack and shone it through the hole into the tunnel. He was amazed by the almost perfect arch it presented, evenly dimpled as if by some intelligent design. The area hewn out of it by the mountain rescue team to recover the body was clearly visible. Shards of broken ice lay in piles all around. Charles Younger's last but one resting place.

'I'm going in,' he told her.

'Be careful,' Addie said, before she realised she wasn't supposed to care. Then added lamely, 'The weight of all that snow on it. It might collapse.'

He scraped away more snow and divested himself of his backpack before stretching the elastic of his headlight around his woollen hat. Addie took his ice axe as he lay on his back and slid himself slowly up into the tunnel. Light from the LED in his torch reflected back at him off every dimpled surface, almost blinding him. He heard more than felt the ice chippings grinding beneath him as he dug in his heels to push himself further inside.

Now he was on a level with the area above him where Younger had been hacked out of the ice. A large concave excavation corresponding very roughly to the shape of a man. Only now did Brodie fully appreciate what a difficult task it must have been to free the corpse from its upside-down grave. He turned his head, directing his light as far as he could above and below, looking for anything that the mountain rescue team might have missed. After all, theirs had been a mission of recovery, not the investigation of a crime scene. And he doubted very much that Robbie's experience would have extended to the latter.

Nothing caught his attention, and he lay still for several minutes, breathing hard, trying to think how Younger's body might have come to be entombed in the ice like this. 'Can you hear me?' he called, and he heard Addie's voice come distantly from the outside.

'Just.'

'What happened here, do you think? I mean, how did he get into the ice?'

'You're the cop.'

'Thank you, that's helpful.'

Several long moments of silence followed, and he wasn't

sure if she was thinking or just ignoring him. Then her voice came again. 'What did the pathologist say happened to him?'

'Someone attacked him, probably on the peak directly above here, and he fell. Broke his neck, fractured his skull and several limbs.'

'Okay,' she said. 'So it was August. No snow on the mountain, except for long-lying patches in the east- and north-facing corries, like this one. Lots of walkers and climbers at that time of year. The body would have been found pretty quickly.'

'So he had to hide it.'

'He must have climbed down and hollowed out a rough grave on the top of the snow patch. Then covered over the body with whatever he'd dug out and more snow from around it. Even in August, temperatures can get down below freezing overnight, but it's warm enough during the day to melt the snow patch just a little. Enough, anyway, for the body to sink down into it, then freeze over again at night. In time, these snow patches become as hard as ice.'

Brodie closed his eyes and saw just how that could happen. A process that would repeat again and again, until the body was subsumed and permanently locked into the ice. But whoever killed him hadn't reckoned on an autumn thaw that would send meltwater running beneath the snow patch, carving out a snow tunnel over the course of several weeks, exposing the body from below.

Almost as if he had spoken his thoughts aloud, she said, 'Then the meltwater exposed him from underneath. Which is when I found him.' And he felt a surging moment of pride in the child he had fathered. There was more of him in her than he had ever realised.

He was about to push himself back out of the tunnel, when the light of his torch caught a fleeting shadow in the ice above him. He stopped and turned his head, raking torchlight among the shattered dimples until he found it again. Something black, the size of a credit card, locked in the ice above his head. He reached an arm towards the tunnel entrance and called, 'Addie, pass me my axe.' And he felt the haft of it pushed into his open hand. He grasped it and pulled it in, lifting it up so that the adze was level with his head. There was very little room to manoeuvre, and it took nearly five minutes of short, repeated hacking movements, ice splintering all around his head, eyes screwed almost shut, to reach the elusive object. He removed his gloves, and with warm fingers melting ice, eventually managed to prise it free. He held it up to the light of his torch, and stared at it, puzzled, for several long moments. It was exactly like a credit card. Black, and completely blank. There was no chip or magnetic strip or engraving of any kind. He frowned, and then it dawned on him what it was.

He thrust his axe back towards Addie and shouted, 'I'm coming out.'

He wriggled his way back out of the tunnel with difficulty, then sat up in the snow piled all around him at the entrance. He held up the card between his thumb and forefinger and Addie looked at it quizzically.

'A credit card?'

'No. It's a keycard with an RFID chip in it.'

'RFID?'

'Radio frequency identification. For opening his car door, maybe. They gave me one just like it for locking and unlocking the eVTOL.'

Addie frowned. 'I don't remember Robbie saying anything about a car.'

'That's because they never found one. But he must have had a vehicle to get here.'

'So where is it?'

He closed his fingers around the card. 'Good question.' He unzipped a pocket in his North Face and slipped it safely inside before getting stiffly back to his feet. He wasn't finished yet with this crime scene and looked up at the steep snow-covered slope of the corrie above him. The fall from the summit must have been a hundred and fifty to two hundred feet. Not a sheer drop, but enough to have inflicted the damage that Sita had found during her autopsy. The top end of the ice tunnel was lost under the recent snowfall, but he wanted to see if he could find it and make his way down through the tunnel from the top end. It would be easier than trying to slide up backwards from below.

He pocketed his headlight and said, 'Wait here. I'm going to try and come through from the top.' And with the help of his ice axe, he began the steep ascent up the corrie. He had covered maybe twelve or thirteen metres before he realized he had lost the profile of the old snow patch. It took several minutes of scouring the slope with trained eyes before he finally resigned himself to the fact that it was probably going to remain buried forever. Or at least until next summer.

He sunk his axe into the snow to begin the process of backing down the way he had come. There was the strangest cracking sound that echoed all around the mountain, and he watched a line from the head of his axe extend left and right across maybe two hundred metres. A vast slab of snow beneath his feet began to slide, and he instantly lost his balance, falling backwards as a sound like the roar of a jet engine filled his ears. Almost the last thing he heard above it, before being submerged by the snow, was Addie screaming.

Now his sense of orientation was gone as once soft snow battered and pummelled him like blocks of concrete, carrying him tumbling down the slope. Without any idea why, he found himself trying to swim through the chaos, arms and legs kicking, as if fighting against the force of giant waves. It seemed to last an eternity. An eternity in which he was strangely conscious of every little thing happening around him. Losing his axe, his gloves, his hat, ragged chunks of ice tearing at his parka, smashing into his face.

And still his instincts were pushing him to swim against the tide of it. Gasping for breath, hearing the grunt of his own voice as the weight of snow forced the air from his lungs. Fragments of blue sky flashed through his field of vision before being lost again in the maelstrom.

Then, almost as quickly as it had begun, it came to an end. He felt himself being dragged down, like a drowning man, and he curled up into a ball, raising both arms in front of his face in the hope of creating an air pocket. Enough oxygen at least to fuel his attempt to get back to the surface.

Silence returned, although his ears were still ringing from the jet-engine roar, and he found himself on his back, one leg apparently clear of the snow above him, the other folded painfully into his chest. His mouth was full of snow, and his teeth hurt from the cold of it. But there was space around his head. He coughed and spat and gasped for air, and heard Addie's muffled voice coming from somewhere distantly above him.

'Dad! Dad!'

Then, unexpected light almost blinded him, and her voice came loud and clear.

'Are you okay?'

And he wondered what primal instincts had been at work that caused him to fight for his life. After all, he was a dead man walking, wasn't he? Wouldn't it have been easier just to surrender and let the avalanche take his life? That would have been an end to it.

But he hadn't wanted to go. Not yet. Not like this. There was stuff that needed to be said first.

He felt Addie's hand grasp his and pull, and he fought to disengage himself from the great chunks of frozen snow that had carried him almost two hundred feet down the corrie. And finally he was free of it, lying on his back, ice-cold air tearing at his lungs, staring up into the sky he had never expected to see again. Everything still seemed to work. He could move both arms, both legs.

Addie was crouched over him, her face etched with concern. 'Fuck,' she said.

He smiled. 'Yeah, fuck!'

And for the first time in more than ten years, he saw her laugh. Something he thought he would never witness again. And he laughed too. And when they stopped laughing, neither of them knew what to say. Each overcome by their own sense of self-consciousness.

Finally she said, 'I thought you were gone.'

He wondered why she would care. But all he said was, 'An avalanche? In November?'

She shrugged. 'Changed days. It was a slab avalanche. Happens when a south-westerly wind blows snow over the summits and into the north-facing slopes. It builds up but doesn't consolidate. Then, if the temperature drops between falls and the snow freezes, like it did late yesterday during the ice storm, the next snowfall will land on the frozen surface and be very unstable.'

He found himself impressed that his little girl even knew such things. He said, 'How come you're okay?'

She smiled. 'I sheltered in the ice tunnel. I really thought it was going to collapse on me. Thankfully it didn't.'

She turned her face to the west. 'Sun's going down. We'd better get off the mountain.'

Without gloves or hat, he felt himself growing colder as Addie helped him down to the foot of the corrie. He remembered that he had left his pack by the entrance to the ice tunnel. A favourite old pack, lost forever. But the keycard for Younger's car was still safe in his pocket, as were his iCom glasses. He put icy fingers to his ears and felt that the iCom earbuds were still there too. A minor miracle.

As they headed west-south-west, down a steep gradient liberally strewn with boulders half-buried in the snow, they were presented with the most stunning of sunset views along the length of the loch below. Snowy peaks glowing pink, flanking a fjord that looked as if it was on fire. Without a word passing between them, they stopped to take it in. Deep, flowing currents and eddies in the waters of the loch burned orange through scarlet, and the last glimpse of the sun slid from view beyond the far mountaintops.

Although the sky was still blue, the first stars were appearing overhead and Brodie thought this is what he would miss the most. That, and knowing that the aching beauty of the country which had nurtured him would still be here long after he had gone. As if his short, unhappy existence in this world had made not one jot of difference. Which, of course, he knew it had not. Addie was his parting gift. The only piece of him that would remain. The only part of him that was any good.

He turned and saw the light of the dying day in her eyes and remembered that she too had a gift to leave the world. He said, 'What's he called? Your boy.'

And he watched the light in her eyes die too. Her jaw clenched. 'That's really none of your business.'

They walked the rest of the way in silence, skirting a deer fence before passing the now derelict and slightly sinister faux-Gothic Mamore Lodge, and down a path that led to a footbridge near the Grey Mare's Tail.

When they got to the village, they stopped at the top of Kearan Road, lights twinkling in the houses around them, but not a soul stirring in the fading evening light. Brodie saw lights on, too, in the medical centre and wondered if Sita might still be there.

He turned to his daughter. 'Thank you,' he said, and meant it.

She inclined her head a little, without meeting his eye, but had nothing to say.

'Well . . . see you, then.'

'Unlikely,' she said, and turned up the path to the door of the police station. Slabs of yellow light fell from several windows, extending across the snow that lay thickly in the garden. There was something warm and welcoming about it. A family. A home. A life. Something he hadn't known for years. He turned wearily to cross the road to the medical centre.

The duty doctor there seemed surprised to see him. 'Dr Roy left some hours ago with the body,' he said. 'Robbie drove them round to the hotel. I saw him coming back a little later.' He frowned at the bruises and grazing on Brodie's face. 'Are you alright?' he asked, and Brodie raised cold fingers to his face, realising that his battered features must reflect the ravages of his brush with death.

'I'm okay,' he said. 'Had a bit of a fall up on the mountain.'

It felt like a long hike back round the road to the International, with heavy legs and a head that seemed likely to split itself open at any moment. Apart from a couple of coffees, he'd had nothing to eat all day. His stomach was growling and he felt almost faint from hunger.

As he walked up through the trees, he saw that there were no lights on in the hotel, its sprawling silhouette standing dark against a starry sky. It was all he could do to drag himself up the front steps and into the hall. He stood for a moment, letting his eyes accustom themselves to the dark, though there was almost no light here to see by, and he fumbled along the wall searching for switches. When eventually he found them, a bleak yellow light filled the hall. At least it was warm in here.

He pushed open the door to the Bothy Bar, but it was empty, brooding in darkness. He turned back into the hall and called out, 'Hello? Anyone at home?', only to receive a resounding silence in response.

Wearily he climbed the stairs to the bedrooms, thinking that perhaps Sita had gone for a lie-down, and irritated by the apparently perpetual absence of the hotel's owner. He knocked on her door, and when there was no response, tried the handle.

It opened into darkness. He found a switch and blinked in the sudden light. Her personal Storm case sat open on the bed, a handful of clothes laid out on the duvet. But there was no sign of her kit, just her torch lying on the bed next to her case. He breathed his annoyance into the empty room. Where the hell was she? And where was all her stuff? Maybe she and Robbie had already loaded her kit on to the eVTOL. He lifted the torch and headed back down the stairs.

He was halfway to the football field when he remembered that Eve was locked, and that he had the only key. But he decided to check on the aircraft anyway. From what he could see by the light of Sita's torch, there were more fresh footprints in the snow, but it wasn't clear whose they were or from which direction they had come.

The eVTOL sat squarely in the middle of the snowy playing field, where she had landed. And was locked, as he had expected. Brodie fished out his RFID card and opened the right-hand door. It was icy inside, and there was nothing of Sita's in evidence. He decided to check the battery level.

'Eve, what is your current state of charge?'

Eve remained stubbornly mute.

'Eve?'

Nothing. Brodie frowned and slipped back out into the snow, closing the door behind him. He crossed to the pavilion to check that the eVTOL was still plugged in. It was, but there were no green lights flashing now on the reader attached to the plug unit. He glanced around. Through the trees, he could see that street lights were still burning in the village, lights twinkling in the windows of dozens of homes huddled around the head of the loch. So there was no power cut.

It was only as he made his way back to the eVTOL that he noticed another sets of prints in the snow. They came from the pavilion, stopping halfway, then turned to make the return trip. There was quite a mess in the snow where they had made that turn. Brodie crouched to shine the light of his torch on the disturbance, and saw that the charging cable had been neatly severed.

CHAPTER THIRTEEN

An icy wind advanced up the loch now, leading a legion of thick black clouds to scrape mountaintops and banish what little light had earlier been offered by the stars. Brodie's parka was zipped up to his throat, its hood pulled around his head, hands thrust deep in his pockets for warmth as he stumbled back around the road to the village.

The cutting of the cable feeding power to the eVTOL was no random act of vandalism. More like a deliberate attempt to stop them from leaving. And he had a dark sense of foreboding about Sita. If she wasn't at the hotel, where was she?

Every muscle in his body was stiffening up from the cold, and from the pummelling he had taken in the avalanche. His eyes felt gritty, his mouth dry, and he could barely swallow.

He hesitated for a long time at the garden gate of the police station. The same warm light as earlier spilled from the same windows. On the walk round, he had tried, unsuccessfully, to make a call to police HQ in Glasgow, but his iCom had been unresponsive, and he was starting to think that it had been damaged in the violence of the avalanche after all. There was nowhere else he could go for help.

The gate creaked as he pushed it open and walked up to the annexe adjoining the house. A blue police sign above the door glowed in the dark. A notice on the door itself read *Knock and Enter*. He did as instructed and the door opened into the warm light of a tiny police office with a public counter and a small waiting area that boasted a couple of scuffed plastic chairs. Robbie sat in a pool of light from an angled desk lamp at a desk on the other side of the counter. A computer screen reflected blue on his face. The clack of his fingers on the keyboard filled the tiny space. He turned, surprised, as the door opened, and the beginnings of a smile vanished quickly, to be replaced by concern. 'What the hell happened to your face? Sir.' The *sir* came almost as an afterthought.

'Didn't Addie tell you?'

He frowned. 'I haven't had a chance to speak to her since she got back. What happened?'

'We got caught in an avalanche.'

'Jesus!' He stood up. 'Is she alright?'

'She's fine. It was me that took the brunt of it.'

Robbie ran a hand back through thick, dark hair. 'I'm sorry I wasn't around when you got back. You probably know already, but all the comms are down. Mobile phones, the police 15G network, the internet. I've been talking to Ballachulish A on short-wave. Old technology, I know, but still good in an emergency. They figure last night's power cut sent a surge down the line that blew the transmitters and the telephone exchange. It's happened before. They'll have sent teams out to get them online again, but who knows how long that'll take.' He pulled a face. 'And there's more snow forecast.' He paused. 'You find anything up there?'

Brodie fumbled in his breast pocket to retrieve the black RFID card and held it up.

Robbie squinted at it. 'What is it?'

'Younger's car key, I figure.'

Robbie was puzzled. 'What car? Brannan didn't think he had one, and there certainly wasn't one in the car park.'

'Then how did he get here?'

Robbie shrugged and made a face. 'There *is* a bus.' But he didn't sound convinced.

Brodie shook his head. 'We can talk about it later. Right now I'm more concerned about Sita.'

A frown furrowed Robbie's brows. 'What about her?'

'I can't find her.'

Robbie advanced to the counter and placed his hands flat on top of it. 'Isn't she at the hotel?'

'No, she's not. Her personal stuff's in her room, but there's no sign of the Storm trunk with her kit and all her samples.'

Robbie extended his hands to either side, perplexed. 'I dropped her off at the hotel this afternoon, along with her kit and everything else. We put Younger in his body bag back in the cold cabinet until you were ready to leave.'

Brodie said, 'Was Brannan there when you dropped her off?'

'No, he wasn't. Why?'

'He hasn't been around all day. And wherever he went, he's still not back.'

Robbie scratched his head. 'Well, he can't have gone far. The road at Glencoe was impassable this morning – though I guess the snow ploughs will have cleared it by now. They've got to keep access to the nuclear plant open at all times.'

Brodie pushed back his hood and opened up his parka at the neck. Now he was too warm, his fingers tingling as they transitioned from ice-cold to blood temperature. 'Something else,' he said. 'Someone cut the cable between the eVTOL and the charging hub at the pavilion. Eve has zero charge.'

'You're kidding!' Robbie's face creased with incomprehension. 'Why would anyone do that?'

'To stop us leaving?'

'But who? And why?'

Brodie shrugged. 'Younger's killer, perhaps. Maybe something he thought Sita would find during the PM. Or maybe he was afraid of something I might discover up on the mountain.' He sighed. 'To be honest, I have no fucking idea.'

Robbie laughed. 'You know what? It was probably some kid who thought it would be a laugh. I know, misplaced sense of humour. But you know what kids are like. And Dr Roy probably walked into the village to get something to eat. There's a couple of pubs that serve food. If Brannan's been gone all day, she wouldn't have got anything at the hotel.'

He snatched his parka and cap from a coat stand and

rounded the counter. 'Listen, I know a guy who can repair or replace your cable first thing tomorrow. But right now, let's go and find Dr Roy. If she's not back at the hotel, we'll do a round of the pubs.'

Brodie felt strangely comforted by the fact that Robbie was taking charge, even though he was very much the junior officer. Brodie was fatigued to the point of exhaustion, and probably not thinking straight. Robbie's suggested explanations for the severed cable and the missing pathologist seemed reasonable, and Brodie was suffused with a welcome sense of relief.

Robbie was reaching for the door handle to go out when the door from the house swung open, and the smell of cooking wafted in. A young boy stood framed in the doorway, a look of consternation writ large all over his face. 'Dad, why can I not get my PlayStation to work?'

'Because the internet's still off, son. Nothing I can do about that.'

But the boy had already lost interest in his non-functioning PlayStation, distracted by the stranger standing by the door with his father. He stared at Brodie with unabashed curiosity. 'Who's this?'

'A police officer from Glasgow, Cameron. Mr Brodie. He's here to help your dad sort out a few problems.'

Brodie was stunned. The skin prickled all over his scalp. The boy had Mel's elfin face, her eyes and nose and mouth. And it was Mel's straight, silky, mouse-brown hair that fell carelessly over his forehead. He had, it seemed, nothing of his grandfather about him, except his name. And that was a shock. Brodie flicked an awkward glance at the boy's father. 'Cameron?'

Robbie seemed embarrassed. 'His mother's choice.'

'What problems?' Cameron said.

'Police problems,' Robbie told him.

'You mean the body Mum found on the hill?'

Robbie was apologetic. He half smiled at the man he now had to reconsider as his father-in-law. 'Village life. Can't keep secrets in a place this size.'

Addie appeared slowly out of the gloom in the hallway behind the boy and slid protective hands over his shoulders, pulling him against her legs. Her eyes were fixed on her father. The silence between them lasted no more than a second or two, but felt like a lifetime to Brodie. He said, 'You called him Cameron.'

And the colour rose almost imperceptibly on her cheeks. 'I always liked the name.'

Brodie attempted a smile. 'It means "crooked nose", apparently. From the Gaelic. Doesn't apply to this handsome lad, though.'

'He gets his good looks from his grandmother.'

Cameron lifted his face towards his mother in surprise. 'I have a granny? Where is she?'

Addie took a moment to compose herself. 'She's in heaven, Cam.'

'And a grampa?'

Addie's eyes never left Brodie's. 'Yes.'

'Where is he?'

'The other place,' she said.

Robbie intervened to break the moment by opening the outside door to let in a gust of ice-cold air and a scattering of snowflakes. 'We'd better be going,' he said. And to Addie, 'I shouldn't be too long.'

Cameron still had fascinated eyes fixed on the man he didn't know was his grandfather. 'What happened to your face?'

Brodie raised self-conscious fingers to his cheek. 'I had a fall.'

'Will you be eating with us tonight?'

Brodie lifted his eyes to meet Addie's. 'I doubt if there'll be time for that, Cameron.' Almost willing her to contradict him.

'No,' Addie said. 'There won't.'

By the time they were on the road back round to the International, the snow had begun to fall in earnest, flying into Robbie's headlights like warp speed in an old *Star Trek* movie. It was wet snow, slapping against the windscreen and gathering in drifts where it was swept aside by the wipers.

Tyre tracks in the hotel drive from earlier in the day had been reduced to mere impressions by the newly falling snow, so Brannan had not yet returned. The only light spilling into the dark came from the hallway beyond the front door. Lights that Brodie had turned on himself just half an hour earlier.

He saw that there were no fresh footprints on the stairs as

they climbed them. In the hall, Robbie called out, 'Dr Roy? Hello, Dr Roy? You back?' He opened the door to the bar, then looked into the dining room, turning on lights as he went. And Brodie remembered Sita's words when they first arrived – was it really only twenty-four hours ago? – *I feel like I've just walked on to the set of* The Shining.

Everything about the place felt just very slightly off. Nothing that Brodie could put his finger on. The brightness of the lights. The smell of damp that lingered on warm air. The worn tartan of the carpet. The flock wallpaper on the stairs. The invasive silence. And perhaps, above all, Brannan himself. His very absence lending him an odd presence. Brodie said, 'She's not here.'

Robbie was scowling. 'I'll check her room.' And he took the stairs to the first floor two at a time. Brodie stood impotently in the hall, melting snow dripping on the carpet. His earlier foreboding had returned. He looked up when Robbie came down again, but the young constable just shook his head. 'I see her stuff still on the bed,' he said. 'We left the big trunk in the room off the kitchen, next to the chill cabinet. Let's just make sure it's still there.'

Brodie followed him into the kitchen. Shadows lurked among the pots and pans dangling above the stainless steel where just that morning they had laid Charles Younger out in a black body bag. Robbie found the light switch, but the disquiet that simmered in the dark was not dispelled by the sudden light reflecting back at them from every shining surface. He pushed open the door into the anteroom and stopped in the door frame, his shadow thrown across the floor and the far wall by the light behind him. 'Jesus.' Brodie barely heard his whispered oath. 'It's not here.'

He reached for a light switch and they both screwed up their eyes against the glare of it.

'We left her case right there next to the cold cabinet. She put her samples in a bag next to the body to keep them cool.' He lifted the misted glass top, and both men found themselves caught in the sightless stare of Sita's dark, dead eyes gazing up at them from the ice-cold interior of the cake cabinet.

CHAPTER FOURTEEN

Brodie sat alone in the dark at the table where he and Sita had exchanged confidences in the bar the night before. To occupy his mind and stop himself from thinking, he had spent five minutes crouched before the hearth, setting and lighting a fire that now sent flickering shadows around the barroom. The crackle of it created the illusion of life beyond the sense of his own faltering existence. But nothing could dispel the deep, deep depression that had settled on him like snow.

Sita's body had been locked in rigor mortis, lying on her back, knees drawn up to her chest, arms folded with her fists at her face, like some bizarre female pugilist. Her killer had clearly experienced difficulty getting her into the cabinet, manhandling her into this strangely unnatural position in order to get the lid shut. It would be hours before rigor wore off and Brodie could remove her from it, to lay her out with dignity.

But perhaps even more bizarrely, the body that hers had replaced was gone. Charles Younger's autopsied corpse in its black body bag had vanished.

Peering in at the dead pathologist, Brodie had seen petechial haemorrhaging around her once beautiful eyes, and a slightly protruding blue-black tongue. There was bruising around her neck. So she had been strangled. It was impossible to tell what other injuries she might have suffered.

Robbie had pulled a chair up to the table in the kitchen and sat with his head in his hands. Face chalk-white. 'I should have stayed with her,' he had said. 'And none of this would have happened.'

But Brodie just shook his head and told him, 'You had no reason to stay, Robbie. No reason to believe she was in danger.'

The young policeman had wanted to remain with him at the hotel, at least until Brannan returned. But Brodie insisted that he go. Robbie's first responsibility was to the safety and well-being of his family. There was nothing more to be done here until communications were restored and they could call for back-up.

It was snowing heavily outside, big wet flakes crushing against the black of the window, blocking any possibility of a view out to the loch, where the lights of the village would be reflected in dark water. He had raided the bar, ripping an almost empty bottle of Glenlivet single malt from an optic to fill his glass. He had been shaking, unable to hold his hands steady in front of him. And the whisky only made him feel nauseous.

For some reason, he couldn't rid himself of the image of the pathologist sitting across from him last night. Her smile, her laughter, her tears. Those dark eyes, and her crinkled black hair drawn back from a handsome face. How unfair it was. After all, he was the one who was dying. The one without a future. Sita had two children who relied on her. And she was still young, with her life lying, in large part, ahead of her. And yet she was the one who lay dead in the kitchen. Crammed unceremoniously into a chill cabinet for cakes and desserts, while he had escaped death just hours earlier in an avalanche. And he couldn't help but feel guilty. Not, for once, as the result of something he had done, or said. But just for being. For surviving.

He should never have come here. Addie had created a life for herself. A family. He had no right to come barging in to ruin yet more lives. He was just a selfish bastard. He was the one who deserved to die. Not poor Sita, leaving her children to the fate of orphans. Tears filled his eyes, and he blinked furiously as they left shiny tracks down his cheeks.

He took a deep breath and screwed his eyes shut, and felt the silence of the International Hotel weigh down on him like a reproach. A log shifted in the fire and sent sparks spiralling up the chimney. A slight blowback produced a small puff of smoke that rose to the ceiling. He could smell it above the damp and the perfume of stale alcohol.

When he opened his eyes again, he saw Addie in his reflection in the glass. He'd had the chance to break his silence when they were on the mountain this afternoon. But he had flunked it, knowing she wasn't ready to listen. Not yet. And even if one day she was, he wondered if he would ever have the courage to tell her the truth.

He closed his eyes again to shut her out, and remember . . .

CHAPTER FIFTEEN

2023

I suppose you might call it an obsession. I couldn't get her out of my mind. She was in my thoughts all day, in my dreams at night. I'm sure that even then, Tiny must have guessed I was smitten. I mean, I never said anything to him, but he knew me well enough to know that I wasn't right. Couldn't concentrate on anything.

I would go home after my shift and watch some streaming movie, and see her face in every actress with long hair. And when I woke up in the morning, I would find myself wondering if she was awake yet, and if he'd hit her the night before. It drove me mad. Until I couldn't stick it any longer.

I had a couple of days off at the end of that week, and I drove across the city to Cranhill on the first afternoon. From a parking spot next to the Cranhill Community Centre on the edge of the park, I could see up Soutra Place to the tower block where she lived with Jardine. I knew which was his car, because I'd checked it out on the police computer earlier in the week. A pillar-box red Mazda MX-5 two-seater roadster. He liked his cars, did Jardine. Worked at a bookie's in town, so he couldn't have been earning that much. But the Mazda was brand new, just a couple of months old, so it was well seen where his financial priorities lay. The year before, he had lost his licence for twelve months for drink driving, so I figured he was being careful not to get into the Mazda if he'd a drink in him. Which must have been hard for an alkie. Because I'd no doubt that's what he was. Just shows what you can do when you've a mind to.

Anyway, it was sitting there in the parking slots for the tower, and I settled down to wait. It was after two when I saw him walking to his car from the entrance, wearing a duffle coat and jeans, and white sneakers. His face was pasty-white beneath that black hair of his, and he'd probably have described the unshaven state of his face as designer stubble. But to me, he looked like he'd just got out of bed.

I heard him pump the accelerator to make the engine growl. I figured he liked that sound, cos he did it several times before putting her into gear and reversing out into Soutra Place at speed. Into first then, and he accelerated hard to the give-way lines at Bellrock Street, barely pausing to look before turning right and powering away up the hill. I almost ducked, afraid he would see me, but he never gave my motor a second glance, and I sat for a good ten minutes after that before turning the key in my ignition and cruising slowly up Soutra Place to park a few slots away from where Jardine left his Mazda.

The lift was working again, and I rode it up to the fifteenth floor accompanied by the smell of urine. I saw the shock in Mel's face when she opened the door to me. And then panic, as she leaned past me to squint down the hall.

'What are you doing here?'

'Just checking that you're okay.'

'Come in,' she said quickly, and took another glance along the hall to make sure nobody had seen me. She shut the door and pressed herself back against it. She was wearing a towelling robe, and her hair hung in wet ropes over her shoulders. I reckoned she was just out the shower and naked under that robe. My mouth was dry and I was as nervous as she was. 'He'll kill you if he comes back and finds you here.'

'He'd not get away with assaulting a police officer a second time.'

She looked me up and down. 'You here officially, then?'

The absence of the uniform kind of gave the game away. 'No.' Wasn't any point in lying about it. 'Anyway, he's gone to work, hasn't he? Won't be back for hours.'

'No guarantee of that.' She pushed past me into the sitting room. It was a good deal tidier than when I'd last seen it. I followed her in and saw her face clearly then in the light from the window. The bruising was mostly gone, just the faintest hint of a scab where he had split her lip. She ran both hands through her wet hair to draw it back from her face. Then stood defiantly, hands on hips, glaring at me. 'Why are you really here?'

'I told you.'

'Why would you even care?'

I hesitated. To tell her would be to make myself hopelessly vulnerable. But I wanted her. Had known it from that first time I set eyes on her. 'You've been on my mind,' I said. 'Every waking minute of every day. When I think about what he did to you, what he might do to you again.'

For a moment, I don't believe she knew quite how to react. But I saw the colour rise on her cheeks, and I wasn't sure whether it was from pleasure or embarrassment. 'He hasn't touched me since that night.'

'Good.'

'So . . .'

'So, what?'

'So, there's no reason for you to worry.'

I reached out to touch her face. I know I shouldn't have, but I didn't have the words, really, to express how I was feeling. She didn't flinch, or move away, her eyes still fixed on me. 'I want to see you, Mel.'

There was something strangely intimate about using her name, as if we knew each other well. I think she felt it too. But she reached up and took my hand away from her face. 'That would be dangerous.'

'I can deal with Jardine.'

'For me,' she said.

And I knew she was right. If Jardine found out I was here, if I were to see her again, it would be Mel he'd take it out on.

I said, 'If you tell me you don't want me to come, I'll walk out that door and you'll never see me again.' Which was wrong of me. I was putting it all on her. Removing any of the responsibility from me.

Still her stare was unwavering. And eventually she said, in a wee small voice, 'He works Tuesday to Saturday, three till ten.'

So every day off, every night shift, I went up in the afternoon. We didn't do anything except talk. She made coffee, and we would sit on the settee together, just blethering. It was funny, I mean we barely knew each other, but within a short time, it's like we had known each other all our lives. Talk was easy, laughter easier. She told me how she'd never known her dad. She figured her mum never really knew who he was.

There'd been a procession of men who'd come and gone at their two-bedroom tenement flat in Tantallon Road. Sometimes they brought Mel presents. Just to shut her up, she thought, to get her out of the way. There had never been any affection. Except from her mum. 'You know when someone loves you,' Mel said. 'They don't even have to tell you. It's just how they are with you. You feel it.' And she glanced at me, a funny little sideways look that made me blush.

And then I went and spoiled the moment by saying, 'Do you feel that with him?' I couldn't even bring myself to use Jardine's name, and she turned her head away quickly, rising then from the settee to head for the kitchen.

'Another coffee?'

I could have bitten my tongue out.

The turning point in our relationship, I guess, came one Tuesday afternoon. I could see immediately that there'd been violence over the weekend. She'd always said that he stayed off the booze during the week, but made up for it Friday and Saturday nights. She'd tried to cover the bruises with make-up, but the damage was still plain to see.

As soon as I got in, I turned her face to the light. 'He fucking hit you again.'

She tried to laugh it off. 'Witch hazel's not working, then?'

It made me so mad. I was physically shaking. If Jardine had been within striking distance in that moment, I'd have fucking killed him. 'Mel, this can't go on.'

She pointed at her face. 'You mean this?' And hesitated. 'Or us?'

I knew there *was* no us. Not really. I mean, we hadn't even kissed, for God's sake. Not that it would have made the slightest difference to Jardine if he knew I'd been coming to the flat. I took her by the shoulders and said, 'I can't let him go on hitting you.'

But she pulled away. 'I can take care of myself, Cammie. I can. I wouldn't have survived this long if I couldn't.'

'Leave him.'

'No.'

'Why not?'

'You *know* why.'

I didn't, not really. Couldn't understand for the life of me why she would stay with a man who beat her. It made no sense. I'd have taken her away from all that shit in a heartbeat. I'm sure she knew that. But he had some kind of hold on her. Something I can't even begin to explain.

She had walked away to the window, and was staring out into the wet afternoon. And suddenly she gasped. 'Oh, my God! He's back! Oh, my God, Cammie, you've got to go.' She turned to face me with real fear in her eyes.

I didn't want to go. I wanted to stay there and have it out with him, but she was very nearly hysterical. In the end I just walked out. Slammed the door behind me. By the time I got to the end of the hall, I could hear the elevator coming. I hesitated for a very long moment. I knew I could take him. But I knew, too, that it could only end badly for me. An off-duty cop beating up the abusive boyfriend of the woman he was in love with. It didn't matter how platonic my relationship with Mel had been so far, it would not play out well.

Reluctantly I slipped into the stairwell as the lift doors slid open. I stood there listening as he went down the hall. The door of the flat opening. Then silence, before I heard raised voices. I closed my eyes and breathed deeply. It took a major effort of will to stop myself from going after him, banging on that door, and beating the crap out of him when he opened it. In the end, I just turned away and began the long descent down the stairs from the fifteenth floor.

I didn't go back for a whole week. I don't know who I was punishing more – me or her. The only winner was Jardine. I felt

Tiny's eyes on me when we were on shift together. You know, kind of . . . appraising. We'd never talked about Mel or Jardine since that first encounter. But somehow he knew. Finally, he said, 'Are you seeing that girl?'

'What girl?'

He made a face. 'Don't come it, Cammie. It's me you're talking to, and you know who I mean.'

I refused to meet his eye. 'No,' I said, and because I wasn't seeing her right then, it didn't feel like a lie.

He gave a little snort of exasperation and turned away, and he never mentioned her again. Until that day in the locker room.

I went back at the end of the week. Absolutely crapping it, in case she told me to sling my hook. I'd watched Jardine roar away in his fucking Mazda, and after riding the pissy elevator to the fifteenth floor, I almost didn't have the courage to knock on her door.

My heart was in my mouth when she opened it. She stood staring at me for a long minute before she nodded towards the open sitting-room door and I went through. I heard the front door shut behind me, and as I turned, she threw her arms around me and buried her face in my chest. I didn't know what to do straight away, I was so taken aback. Then I put my arms around her, too, and felt her whole body quivering. We'd never been this close before. I'd never felt her body against mine. It was electric.

'I thought you'd given up on me,' she said.

I made her stand back from me, and took her head in my hands, smearing away her tears with my thumbs. 'I'll never give up on you, Mel. Never.' And I never did.

She tried to control her breathing between sobs. 'You can't come back here, though. You can't. I'm sure he suspects. I'll meet you somewhere. Somewhere in town where he's never likely to see us together.'

Which is when we started meeting at the Cafe21 in Merchant City. It was one of those cafés that was all wood and brick and steel on the inside, and glass and marble outside. Typical Glasgow, they put cane tables and chairs out on the pavement, more in hope that it wouldn't rain than in any expectation of sunshine. You could try all you like to pretend it was Paris, but in Glasgow that never really washed.

The Merchant City was one of the oldest parts of the town. It's where all the wealthy merchants from the days of empire had their warehouses, shipping tobacco and sugar and tea. Then later it was home to the city's fruit and vegetable and cheese markets. By the time me and Mel were meeting at the Cafe21, it had become the in-place for posh folk who didn't mind spending a bit of cash in the boutiques and gourmet restaurants. Not a place Jardine or any of his cohorts would ever be seen dead in.

We always took a table up on the mezzanine. Mostly we just had cappuccinos, but sometimes we would get stuff off the menu if we were hungry, or they were busy and we wanted to keep our table. They had wraps, and toasties, and nachos, as well as pizzas and stuff. It was okay, but it wasn't cheap.

I didn't care, though. I was with Mel, and we weren't worrying every minute of our time together if Jardine was going to come back unexpectedly and catch us.

I remember those days with such an aching fondness. Away from that flat in Soutra Place, she was a different person. Relaxed, so quick to laugh, interested in every little thing about me.

I told her how my mum died when I was young, and really it was my dad who brought me up, in a single end in Clydebank. He'd been an apprentice welder in the shipyards when he was young. Though, even then, there weren't that many shipbuilders left on the Clyde, and when the yard where he worked closed down, it had been almost impossible for him to find another job.

'There was a time,' I said, 'when he really thought about us emigrating to Australia.'

Quite impulsively, she reached across the table to grab my hand. 'Oh, I'm glad you didn't.' And the touch of her hand on mine suffused me with such warmth, I find it hard to describe. I put my other hand over hers and hoped, somehow, that everything I was feeling would be transmitted from my heart to hers through our touching hands.

I laughed. 'Well, he'd never have taken me hillwalking if we had. I don't know if there are any mountains in Australia, but I'd have missed bagging all those Munros.'

She frowned. 'Munros?'

And I explained to her what a Munro was, and that they were named after some toff called Munro who'd made a list of them all.

'I'll take you with me some time,' I said. 'When you get up there among the peaks, it's like you're on top of the world. Puts everything into perspective, and you realise how small your problems are by comparison.'

'I'd like that,' she said, then laughed. 'But I'd probably have to go into training for six months first.'

I'd been meeting her at the Cafe21 for maybe six months. Sometimes there was bruising. Sometimes there wasn't. I never mentioned it when there was. And she never once talked about her life with Jardine. It was like, you know, a cat that hides its head beneath a cushion and thinks if it can't see you, you can't see it. We were just pretending we had a life together. If we didn't talk about the rest of it, then it didn't exist.

It was one early spring day when we met in the late afternoon and she told me she wouldn't have to be home till late that night. Jardine thought she was going on a girls' night out, and wasn't expecting her to be there when he got in from work.

I remember thinking I could take her to a movie, or out for a meal somewhere. Maybe even take in a show in town. I made a couple of suggestions, and she sat there looking at her hands folded in her lap. Then she raised her eyes to mine and said, 'Maybe we could just go to your place.'

My heart kind of thundered around in my chest for a minute before pushing up into my throat and damn near choking me. I knew this was what they called a watershed moment. The direction of our relationship was about to change course. And if we went with the flow, there would be no way back.

We took a taxi to my place at Maryhill. Sitting in the back saying nothing. But we held hands for the first time. I mean, it was really no big deal. But it kind of was. I was so nervous. It wasn't like I'd never slept with a girl before. There'd been a few. But this was different. I wanted it to be amazing. The best ever. And I was scared it wouldn't be.

I thought maybe she felt that way, too. But when we got back to the flat, she was all over me the minute the door was closed. Hungry for me, like she hadn't eaten in a month. And all my fears fell away, like the trail of clothes we left on the floor on the way to the bedroom. Jesus! And it *was* amazing. Better than I could ever have hoped. Better than I could ever have imagined. I was so lost in her, so blind to the future, that I couldn't see how impossible it all would become.

Addie was conceived that night, though I didn't know that till much later. But I told Mel for the first time that I loved her. First time, actually, that I ever told anyone that. I'd never had the faintest idea what love was, or how it was supposed to feel. But I did now, even if I couldn't put it into words.

We lay together afterwards, till it got dark and street lamps sent their orange light through the window in long boxes deformed by the tangle of quilt on the bed. We said nothing in all that time, till finally it was Mel who broke the silence. And she said, quite simply, 'Cammie, I'm scared.'

It was strange how our meetings at the Cafe21 were never quite the same after that. Like they weren't enough now. We both wanted more and better, but the opportunity simply wasn't there. I'd never had any control over when, or for how long, we could meet. And before the night at Maryhill, I'd been able to thole that. Just. Now, I couldn't. And while everything had changed for us, really nothing had, and I was just about demented.

Then fate intervened, in a way that neither of us could have foreseen. I'd met Mel briefly at Merchant City that Saturday afternoon. She'd been depressed. The weekend always did that to her. Jardine had been drinking the night before, and he'd be drinking again tonight. She always faced it with a kind of stoic endurance, but I was finding it harder and harder to take. I tried again to persuade her to leave him, and the shutters came down, just as they always did. She wouldn't even discuss it. We had words. I gave her an ultimatum. As I had done several times before. But she knew they were just empty words. That I'd never give up on her. Because I'd told her that, hadn't I? I didn't ever want to lose her. And she knew it.

So we parted on bad terms that night, and I was feeling particularly low when I started on the early shift Sunday morning. I was getting changed in the locker room when Tiny came in and sat down beside me on the bench. He looked grim. 'Got some news for you, pal.'

I couldn't conceive of news, good or bad, being of any interest at all to me right then. I just grunted and said, 'That right?' and bent over to tie my laces.

Tiny said, 'I know you've been seeing that girl.' And when I straightened up to deny it, he put a hand on my arm and said, 'I *know* you have, mate. And I'm figuring the only reason you're not an item is cos she won't leave him. Am I right?'

I tried to stare him down, but I couldn't, and finally went back to tying my shoelaces.

'He's in the slammer.'

And I straightened up again so fast I almost slid off the bench. 'Who?'

'Jardine.'

Now I was alarmed. 'What did he do to her?'

'Nothing. He was out in that flash red sports car last night with a bucketload of booze in him. Over on the south side. Mosspark Boulevard. Doing upwards of ninety by all accounts. Lost control and slammed into an oncoming vehicle. A family SUV with a mother and two kids in it.' He paused and pressed his lips together in a grim line. 'All dead.'

'And Jardine?'

'A few bumps and bruises. Always the way of it, isn't it?' He shook his head. 'Fucking horrible thing to happen. But, mate, he's going down for a long time.'

To be honest, I couldn't think of anything other than that

poor woman and her two kids being dead, and that cunt still walking this earth. Maybe if I'd taken him on. Maybe if I'd given him the hiding he deserved and taken Mel away, things would have turned out different.

I put my elbows on my thighs and buried my face in my hands.

Tiny was concerned. 'You alright, Cammie?'

I sat up and shook my head. 'No. I should have fucking killed him when I had the chance.'

'Then you'd be the one getting sent down.' He put an arm around my shoulder. 'Mate, things just happen. Some of them you can control, most of them you can't. I don't know how things'll be for you and that lassie now. And I have to admit I've never really understood what it is you see in her. Only you know that. But one thing's for sure – Jardine'll no' be an issue now.'

Jardine's case wasn't in the system for as long as you might have expected. He pleaded guilty at his first appearance, when it is usual to make no plea or declaration. Maybe his lawyers told him that he'd get a lighter sentence if he didn't put the court to the trouble and cost of a trial. The Scottish Government had just passed a law increasing the sentence for causing death by careless driving under the influence of drink or drugs to life imprisonment. Or maybe Jardine just wanted it over and done with. At any rate, I was in the courtroom the day he was sentenced, just to show support for Mel. Even though we sat well apart. There were some unsavoury relatives of Jardine's on the public benches, too. A hard-faced sister. An aunt and a couple of sketchy cousins. As well as what I would have described as several acquaintances of dubious character. Other than that, the public benches were largely empty. Nobody was much interested in the fate of Lee Alexander Jardine.

He stood in the dock flanked by a couple of uniformed officers. He seemed unrepentant, and I figured that the social work reports that the judge had received probably made pretty grim reading. When asked if he had anything to say, he just shook his head.

'Speak up for the record,' the judge told him.

'No comment, Your Honour.'

Twenty years was the decision. Some in the court might have thought that harsh. Personally, I thought life would have been too fucking short.

Jardine himself showed no emotion. Just before they led him down to the cells, he turned and scanned the benches behind him. His gaze fell on me, and lingered there for a moment. I had no idea how much he knew about me and Mel, if anything, but for those brief seconds I felt bathed in his hatred, before he glanced at Mel, and a sick, sad smile washed momentarily across his face.

I was parked on the other side of the river, and sat waiting there for Mel for nearly half an hour after the sentencing was over. I was beginning to think she'd stood me up when I saw her trauchling across the Albert Bridge. She looked like she had the weight of the world on her shoulders, and there was something infinitely sad about the way she held herself.

I'd seen her only a couple of times, and briefly, since the crash. I had no idea what she'd had to deal with, what kind of relationship she had with Jardine's relatives, or the friends who had probably come to Soutra Place to offer comfort and who knew what else. I was sure she must have been to visit Jardine in the remand wing at Barlinnie. All of which meant I had no real idea where we stood now. And, in all likelihood, no say in where we went from here.

She slipped into the passenger seat and sat gazing out the windscreen, back across the river towards the High Court. I couldn't even bring myself to speak. Afraid that, whatever I said, it would be the wrong thing. I saw a single tear track its way slowly down her cheek. She said, 'My mother had a weakness for the horses. Had an account with the local bookie. Sometimes she'd place her bet by phone. But more often than not, she went to the bookie's in person. They always made a fuss of her there. And quite often she'd take me. Showing me off. She was still a good-looking woman then and liked folk to think we were sisters. I was probably only fifteen when I first met Lee there.'

She glanced at me. She knew I didn't like her to talk about him, so this was the first time I'd heard how they met.

'He had a sort of wide-boy charm, you know. My mum was a good bit older than him, of course, but I think he quite fancied her. He could make her laugh, and he worked hard at it.' She paused, lost in wordless recollection. 'I was eighteen when my mum OD'd. The guys from the bookie's all came to the funeral, and it was Lee who took me home after.' She shrugged. 'That was the start of it, I suppose.'

And I had the first inkling of what it was that drew her to him like a moth to the flame. He was more than a lover. He was the father figure she'd never known. No matter how abusive he got, he was some kind of anchor. Gave her life shape and stability, even if the only predictability in it was that he would get drunk every weekend and raise his fists to her. I remembered her telling me that first time we met how he'd bring her flowers and chocolates, and take her out to nice restaurants after the violence. His way of showing penitence for the way he was when he drank.

I said, 'It's over, Mel. You're free of him.'

She turned and looked at me. 'Free?'

'To start a new life. Build a future that doesn't include violence and abuse.'

She nodded and wiped away that single tear. 'I'm pregnant, Cam.'

I was so shocked, at first I couldn't even speak. I was scared to ask, but I had to know. 'Is it . . . mine?'

She nodded.

'How can you be sure?'

And she raised her voice, just a little, to lend it certainty. 'Because I am.' She looked at me so directly then that I very nearly had to look away. 'That new future you see for me, Cam: it won't be anything if it doesn't include the father of my child.' As if she thought for one moment that I would let her go. Either of them.

It took her less than a month to settle affairs at Soutra Place and move in with me at Maryhill. Free of Jardine, she seemed like a different person, and there was no impediment to our relationship being whatever we wanted it to be.

Some nights we sat up in bed watching TV, eating ice cream from a local deli and drinking port. Well, I drank the port. Mel wouldn't touch alcohol till after the birth. We made love at any time of the day or night. Whenever the notion took us.

She wasn't much of a cook, so we lived mostly on carry-out pizza, or Indian or Chinese. We ate out a lot, and she made me take her to the ballet at the Theatre Royal. She'd always wanted to go, she said. I suppose all little girls are drawn to the ballet for some reason. We sat in the front seats. Close enough to hear the thumping and grunting, and smell the elephant odour of straining bodies sweating in nylon. She loved it. I hated it. And we laughed about it long and hard in the pub afterwards.

Mel was presenting quite a bump when we got married six months into her pregnancy. It was a dead simple affair at the registry office in Martha Street. Tiny was my best man. His Sheila was Mel's best maid. They met for the first time on the street outside. Witnesses, the registrar called them. And they were, indeed, the only folk to witness the short ceremony. We had an awkward biryani afterwards at their favourite Indian in Shawlands, and me and Mel were just happy to get home and carefully consummate our new-found status as man and wife. God, how I loved that girl!

Three months later, Addie came into our lives and we moved to a semi in a south-side suburb with a wee pocket-handkerchief square of garden at the back. I built Addie a swing, and a see-saw. I taught her to ride a bike, how to swim. I adored that wee girl, and she loved her daddy.

In the years that followed, Tiny and I sat and passed all our exams and moved up the ladder. CID, plain clothes, working now out of the new HQ at Pacific Quay. Tiny and I were still pals, though me and Mel hardly ever saw him and Sheila as a couple. Sheila still didn't like me much, and the feeling was still mutual. And I'm sure she disapproved of Mel.

Where do the years go? I mean, it seemed like no time since me and Mel were meeting secretly at the Cafe21. And now Addie was in her teens, all hormonal and awkward and doing her best to piss me off at every turn. I think, maybe, she was closer to her mum in those years. But we were a family, even if Mel never did get pregnant again, and there was a lot of love there. We'd moved into a red sandstone semi in Pollokshields by then, and Addie had not long turned seventeen the day I logged in at Pacific Quay to find Tiny sitting at my desk in the detectives' office. He was swivelling back and forth in my chair, legs akimbo, sucking on the rim of a disposable coffee cup.

In my usual polite way, I told him to fuck off out of my chair.

But he didn't budge, just sat there staring at me thoughtfully. Then he said, 'You heard?'

'Heard what?'

He hesitated for just a moment. 'Lee Alexander Jardine is out on licence.'

CHAPTER SIXTEEN

2051

A muffled thud from somewhere deep in the hotel startled him.

The fire he had lit earlier was a faint glow as the last of its embers turned to ash. The snow outside was still blowing hard against the window, even wetter now, and running down it in sleety rivulets. A few moments before, he had forced himself to drain his glass. There was no real escape in the drink, he knew that. There never had been. He had learned long ago that no matter how much you drank, everything that made you seek refuge in it was still there in the morning, when you woke with a splitting head and a mouth so dry it was an effort to peel your tongue off the roof of it. But as his old history teacher had been fond of saying, the only thing we learn from history is that we never learn from history.

Now he sat up, heart pounding, blinking hard to try and clear the fog of grief and alcohol from his brain. There was someone else in the hotel.

Brodie got to his feet and crossed to the fireplace. From the selection of fire irons, he picked out a wrought-iron poker with a viciously curling log hook. He hefted it in his hand to feel the weight of it. It would do some damage to anyone on the receiving end. Then he turned to face the door.

He had been in and out of that door to the bar several times over the last twenty-four hours, and never noticed the noise of its hinges. Now they screamed in the dark, like poor Sita's lost ghost. He was sure he had left the lights on in the hall. But it was pitch-dark beyond the bar now. He stepped cautiously on to the tartan carpet and waited for his eyes to accustom themselves to the lack of light, his breath coming in short, sharp rasps.

Another noise set him on edge. A clatter this time. And it seemed much closer. Beneath the door leading towards the rear of the hotel, he now saw the faintest line of light. He stood listening intently, but could hear nothing above the rush of blood in his ears. As he advanced towards the door, there was more clattering beyond it. He pushed it open, and saw a hard line of bright, clear light beneath the kitchen door. A shadow moved about behind it, breaking the line of light. Brodie braced himself and ran at the door. It flung itself open with the force of his shoulder and he was momentarily blinded by the kitchen lights.

Brannan turned, startled, from the stove. Steam rose from a pot on the rings. His eyes were wide and frightened as he took in the figure of Brodie brandishing a poker. He raised a hand, as if that might protect him from the blow if Brodie were to attack. 'Jesus Christ, Mr Brodie! What are you doing? You just about gave me a heart attack.'

Brodie stood staring at him, half in relief, half in anger. 'Where the fuck have you been, Brannan?'

If anything, Brannan was even more startled by his tone. 'What do you mean?'

'You have guests in your hotel and you haven't been here all day. And there's been murder committed under your roof. Where the fuck have you been?' It was in danger of becoming a refrain.

Whatever was in the pan on Brannan's stove began to boil over and he turned quickly to remove it from the ring. He was pink-faced and flustered. 'I had to go to a funeral this morning. Other side of Ballachulish. There was a meal. And then, you know, things carried on into the afternoon. The wake and everything.' He paused, as if Brodie's words had only now fully sunk in. 'Murder?'

He cowered as Brodie strode across the kitchen and grabbed him by the arm, propelling him towards the anteroom door and kicking it open. Brannan staggered as Brodie pushed him inside. He waved his poker towards the cold cabinet. 'Open it!'

There was something very close to panic in Brannan's eyes. 'Why?'

'Fucking open it!' Brodie's voice resounded around the enclosed space as he pushed Brannan towards the cabinet.

Brannan steadied himself, breathing rapidly, and lifted the

lid, almost afraid to look inside. When he did, he emitted a strange half-strangled cry, and staggered backwards, as if pushed, crashing into the shelves on the wall behind him, sending cans of peas and asparagus and jars of preserve clattering away across the floor in the semi-dark. He glanced towards Brodie, naked fear replacing panic. He was so breathless his voice came in a whisper. 'It's Dr Roy!'

Brodie took a step towards him and pushed the point of his poker into Brannan's throat, the log hook curling around the line of his neck. 'There was no funeral on the other side of Ballachulish this morning,' he said. 'The road was closed because of snow.'

Brannan recoiled from Brodie's breath in his face. 'You . . . you've been drinking,' he said.

'Your finest Glenlivet. And even if I'd put away a whole bottle of it, Sita would still be dead. So now you're going to tell me where the fuck you were, or so help me, I'll rip your fucking throat out.'

And Brannan had no doubt that he meant it. 'Okay, okay.' And very gingerly, with thumb and forefinger, he took the end of the poker and drew it away from his neck. 'Can we go back to the kitchen, please?'

Brodie glared at him a moment longer, then stepped away to allow Brannan to pass before following him into the kitchen. Brannan stood under the glare of the lights, trying to catch his breath and his composure before turning to face Brodie.

'I could do with a drink.'

'So could I.'

Brannan canted his head quizzically and said, 'Do you not think you've had enough?'

'There's never enough,' Brodie growled.

Brannan said, 'When did you last eat?'

'What do you care?'

'I'm hungry. I'll make us both some supper.'

There had been a residual warmth in the bar from Brodie's fire before Brannan took the poker carefully from Brodie's grasp and stoked the embers. The couple of logs he'd thrown in were crackling now. They sat at a table together, forking mouthfuls of spaghetti carbonara into their faces. Brannan's panic had subsided, but his face was sheet-white, and his hand trembled as he wielded his fork. He was reluctant to meet Brodie's hostile eye.

'I went to the 3D houses to talk to Charles Younger's source at Ballachulish A,' he said.

Brodie's eyes crinkled in confusion. 'Source?'

'He wasn't there. He was on night shift at the plant. And his wife said he'd be back early afternoon, once the road was cleared. She told me I could wait, and she offered me lunch.'

Brodie shook his head. 'What do you mean, *source*?'

'Joe Jackson. He's a reactor operator at the nuclear power plant. He'd not long started at the plant and was living at the hotel when I bought it. Then they allocated him one of the 3D houses this autumn, and he was able to bring his family up to join him. I got to know him quite well. Nice guy.'

Brodie banged his fist on the table and Brannan jumped. 'You're not making any sense to me, Brannan.'

Brannan swallowed over a mouthful of carbonara. 'When Charles Younger was staying here, I saw the two of them in the bar one night. Huddled together in a corner just over there.' He waved vaguely towards a dark corner beyond the pool table. 'It was busy. I didn't think anything of it. But then they were there again the next night. And the next. It was all tourists here in August, so no one knew who they were. But I just happened to mention one day when Joe was leaving for work that he seemed pretty friendly with the journalist. It was like I'd stuck a firework up his arse. He physically jumped. Told me he didn't know him at all. They'd had a few drinks together, nothing more.'

'And?'

'Well, I thought it was a pretty strange sort of reaction. I mean, a couple of guys having a drink together in the bar. What's to get jumpy about?' He drained his glass and refilled it as he spoke. 'Anyway, when Younger went missing, Joe was all in a panic. Pulled me aside and pleaded with me not to mention to Robbie that he and Younger had been drinking together. He was really spooked.'

'And it never occurred to you that Joe might have been responsible for Younger's disappearance?'

'Hell, no! He's not that kind of guy. He's more . . . cerebral, if you know what I mean. No way would he have been involved in whatever happened to Younger.'

'Well, let me tell you what happened to Younger. He was murdered, Brannan. Someone attacked and assaulted him at the summit of Binnein Mòr and pushed him into the Corrie of the Two Lochans, where he broke his neck in the fall. And something Sita found during the post-mortem made her a target, too.'

'Well, whoever killed Dr Roy, it couldn't have been Joe. He didn't get back from the plant till after two, and I've been with him all afternoon and half the night.'

'Why?'

Brannan sighed deeply. 'Trying to persuade him to talk to you.'

'Why?' Brodie was insistent.

'Because he's probably got a good idea why Younger disappeared.' He paused and rephrased. 'Why Younger was murdered. But we didn't know that this afternoon.'

'Why would he know anything about Younger's disappearance? And why would you think he did?'

'Well, like I said, it was obvious that Joe was some kind of source.'

'Source of what?'

'Information.'

Brodie was losing patience. 'Information about what, for God's sake?' His raised voice echoed around the bar.

Brannan shrugged hopelessly. 'I don't know. About the plant, I suppose.'

'Ballachulish A?'

'Well, what else would he know about?'

'So he was some kind of whistle-blower?'

'I wouldn't know. I really wouldn't.'

'So what did he say during all those hours you were with him today?'

'Just that he didn't want to get involved. He was scared. Rabbiting on about the safety of his family. His future. I just about got his whole life story.' He took a mouthful of spaghetti and chewed on it for a few moments. 'Look, Mr Brodie, I went out on a limb here. I don't want my hotel dragged into this. But ever since they found Younger's body, I knew there had to be more to it. That Joe must know something. And I'm sure he does. But he's just so . . . so scared.'

Brodie finished the last of his carbonara, then leaned across the table towards Brannan. His voice was low and dangerous. 'Well, you tell your friend that if he doesn't talk to me and come clean, I'll be going after him. Hard. Okay?'

'Okay, okay. I'll talk to him again. First thing tomorrow. I promise.'

CHAPTER SEVENTEEN

A soft knocking at the door slowly penetrated the layers of fatigue that had wrapped themselves around his consciousness. Like a man rising from the deep, he broke the surface and opened his eyes to be greeted by a grey light that filled the room. He turned his head a little to the side. Through the window he could see low-lying clouds, bruised and battered, hanging from a turbulent sky, and big white flakes of snow drifting down beyond the glass.

For the second night he had slept in his clothes. He scratched the whiskers that bristled across his unshaven face and blinked the sleep from his eyes.

The knocking at the door came again. Louder this time. And a voice from the other side of it called, 'Mr Brodie?' A voice he didn't recognise. He sat up too quickly and felt momentarily giddy.

'Just a minute,' he growled.

Slowly he swung his legs around to put his feet on the floor and stood up. He crossed to the sink and sluiced his face with cold water, then lifted his head to see the wreck of the man he'd become staring back at him from the mirror with blood-shot eyes.

He opened the door to find a short, thick-set man with a silvering beard standing in the hall. A blue fleece was dragged over a green chequered shirt. He was almost completely bald, and clutched a patterned woollen hat in his hands. Brodie could see the shock in wide-set blue eyes as he took in the state of the police officer who opened the door. Behind his embarrassment, a smile returned to friendly eyes. 'Mr Brodie.' Not a question. He thrust his right hand towards the policeman. 'Calum McLeish.' Brodie shook it. 'I'm on the mountain rescue team with Robbie. Electrical engineer. Work up at the hydro plant. Robbie asked me to come over to see if I could repair a severed charging cable.'

Brodie had forgotten all about it. 'Give me two minutes,' he said, and closed the door in the other man's face.

They drove around to the football pitch in McLeish's dark blue pickup truck, making fresh tracks in thick, wet snow. The sky hung low, still spitting snowflakes into the chill morning air and obscuring the peaks that surrounded them. Through snow-laden trees, he saw the houses of the village grouped around the head of the loch, reflecting in slate-grey water. There was barely a breath of wind to disturb its mirrored surface. And nary a sign of life.

Brodie had tried his iCom before leaving the hotel, but there was still no signal. McLeish had watched, fascinated. 'New comm kit?' he said.

Brodie nodded.

'Very cool.'

But Brodie shook his head. 'Not worth a damn if there's no signal.'

When they reached the eVTOL, McLeish jumped down into the snow, his breath billowing about his head as he pulled on his waterproof jacket. He leaned into the back of the truck to retrieve a toolkit from the flatbed and heaved it up over the side wall. Then he stood gazing admiringly at Eve. 'She's a fine beast,' he said. 'Not ridden in one of those before. Smooth, is it?'

'Unless you're flying through an ice storm.'

McLeish grinned. 'Aye, well, that wouldn't be very comfortable in anything airborne. Where's the cable?'

It was buried under the new snow. Brodie grabbed the end at the eVTOL and started pulling it up as they headed towards the pavilion. 'Don't these things usually have contactless charging?' McLeish said.

Brodie grunted, barely able to keep a civil tongue in his head. 'Do you see a contactless charger around here?'

But McLeish maintained his good humour. 'Good point.'

Finally the cut end of the cable pulled itself free from the snow, and Brodie crouched down to search for the other end.

'Is it still plugged in?' McLeish said.

'It was the last time I looked.'

'I'll go and unplug it, then. Be unfortunate if we both ended up fried for breakfast.'

The very thought of breakfast made Brodie heave, and he stood up, breathing deeply, as McLeish walked over to the pavilion to unplug the cable. When he came back and examined the cut ends, he shook his head.

'Someone took their life in their hands cutting through this. Must have had well-insulated wire-cutters.' He looked up at Brodie standing over him. 'Why didn't he just unplug it?'

'Presumably so it couldn't just be plugged in again.'

'Aye, right enough, I suppose, if the object of the exercise was to stop the battery from charging . . .' He opened his toolbox, set in the snow beside him. 'I can do a temporary repair to get it charging. But it'll need to be handled with care, and best keep it clear of the snow. Don't want water getting in and shorting the thing.'

Brodie stood watching as McLeish stripped back the cable from either side of the cut ends to prepare the wires for reconnection. 'How long have you been on the mountain rescue team?' he said.

'Since I was a teenager, Mr Brodie. My dad was the team leader then. Taught me everything there was to know about the mountains.'

'So you're from the village?'

'Born and bred. Nowhere else I would rather live. Especially in this day and age. I've seen some changes in the world in my time, as I'm sure you have, too. Most of them for the worse.'

Brodie nodded. 'You were part of the team that brought down the body, then?'

McLeish looked up from his repair. 'I was that, Mr Brodie. I've brought a few bodies down from the mountains over the years, but never saw anything like that before. What a helluva job it was getting him out of the ice.'

'I don't suppose you'd have any thoughts about what he was doing up there?'

McLeish shook his head. 'Not a one. And from everything I hear, he was a rank novice. I mean, superficially he had the right gear and everything, but from what I could see, it was all brand new. His boots, for example. No wear on them at all.'

Brodie crouched down beside him and watched him work the wires for several minutes, before he said, 'If you were going to go up the mountain and wanted to leave your car as close as possible to the start of the climb, where would that be?'

McLeish looked up from his work and thought about it. 'Depends which way you were going to go up. I mean, there's an easy way, and a hard way. But the easy way's a long trek and the hard way's the quickest way down.'

And Brodie thought, that's how he and Addie had done it. The long way up, the fast way down. Faster than he'd have liked. 'Do you have a map?'

'Aye, in the pickup.'

'Could you show me?'

'No problem.'

They returned to the truck and McLeish retrieved a map from the glovebox, spreading it out on the passenger seat.

'Here,' he said, pointing out the route that Brodie and Addie had followed up into the trees from the Grey Mare's car park. 'Easy way up.' Then traced a finger along the route that father and daughter had taken to come down off the mountain. 'Hard way.' He looked at Brodie. 'I take it we're talking about Mr Younger?'

Brodie nodded acknowledgement.

'Well, then, as a novice, it would make sense for him to take the easy way up.'

'Yes, that's what Archie McKay said.'

'McKay? You've been talking to that blowhard, have you?'

'So had Mr Younger, apparently.'

McLeish raised an eyebrow in surprise. 'Had he now? Archie certainly kept that one to himself. And what advice did the old bugger give him?'

'Same as you, Mr McLeish. But apparently Younger told him he didn't have time to take the long way round. Being a novice, maybe he didn't realise just what a tough climb the short route would be.'

McLeish snorted. 'Aye, well, he wouldn't be the first to make that mistake.'

Back at the hotel, Brodie forced himself to eat the two fried eggs with Lorne sausage that Brannan had prepared for breakfast, and drank nearly a whole carton of orange juice.

Then he returned to his room to wash and change, and gazing at his ravaged face in the mirror, decided that a shave might make him feel better. He had just laced up his climbing boots and was pulling on his North Face when there was a knock at the door. He thought it was probably Brannan with news of Joe Jackson. 'Yeah?'

The door opened and Addie stood framed in the doorway. For a young woman who always seemed so sure of everything, in that moment she looked very uncertain. He straightened up and gazed at her with an ache of regret somewhere deep inside.

She said, 'Robbie told me about Dr Roy. I wanted to come last night, but he said you might not be very . . . receptive.'

He forced a smile. 'He might have been right.' He paused. 'Why would you want to come anyway?'

He saw a tiny shrug of her shoulders. 'I don't know. Seemed like such a shitty thing. I suppose I just wanted to say sorry.'

'It's not your fault.'

'No. I mean . . . just sorry for all that's happened. To Dr Roy and everything.'

He looked away. 'She was a nice lady. Lost her husband a while back. Had two young kids, too.' He felt himself choking up again. 'She didn't deserve that.' He zipped up his parka. 'I could use your help if you have time.'

'With what?'

'I'm going to look for Younger's car.'

'You know where it is?'

'No. But I figure he probably drove it as close to the start of the climb as he could get. If he took the route that you and I did, that would be the Grey Mare's car park. And it's not there. So . . .'

'He would have taken the old military road,' Addie said, 'if he was going the other way.' She paused. 'But it would be crazy for a beginner to attempt that route up the mountain.'

'Aye,' Brodie said. 'Just the sort of thing someone who didn't know any better might do. But to be fair to him, he did actually get to the summit.'

'If he'd left his car somewhere on the road, it would have been seen.'

'Maybe.' He picked McLeish's map off the bed and traced the line of the old military road. 'It looks like there's some kind of off-road area here.' He stabbed a finger at it. It was well above the stream that ran down through the trees and eventually tumbled over the rocks at Grey Mare's Waterfall. 'I don't want to have to go the long way round and follow the road up to it. Can you guide me through the trees from below?'

She sighed, pressing her lips together. He knew that look. That forced concentration when she was undecided about something.

He said, 'One way or the other, I'm going. But it might make things easier if . . .' He let his voice trail away.

She gave him a hard look with her mother's eyes. 'Since you never answered me the last time, I'm going to ask again. Why are you here, Dad? Really? It's not for this, is it?' She waved an arm vaguely around herself. 'Some missing person. A murder enquiry. I mean, do your bosses even know I'm your daughter?'

Grudgingly he shook his head.

'I didn't think so.'

'I volunteered.'

She forced a breath of deep frustration through her lips. 'Why?'

'There's stuff I have to tell you.'

She shook her head vigorously. 'I don't want to hear it.'

'I know you don't. But you have to.'

'I don't have to hear anything from you.'

'Yes, you do!' His suddenly raised voice startled her. 'Addie, I've held things inside of me for the last ten years. Mostly guilt.'

'And now you want to offload it on to me.'

'No.' He was shaking his head slowly, internalising some buried pain. 'I'll take my guilt with me to the grave.' He turned penetrating blue eyes on her. Eyes that were filled with something she had never seen before. Something she couldn't define. But they were almost chilling in the way they violated all her outer defences. 'There are things I need to tell you. Things you need to know.' He hesitated. 'And I need to tell you now, because . . .' He couldn't bring himself to say it.

Something in his desperation caused her heart to skip a beat. 'Because what?' And perhaps because she feared the answer, she provided one for herself. 'Because it'll make you feel better?'

All the fight went out of him. She saw him go limp, and his gaze drifted away to some far-off place. 'Because if I don't tell you now, I never will, and you'll never know the truth.'

Now she really was afraid to ask. Her voice was very small. 'Why?'

His eyes flickered up to meet hers and she saw the defeat in them. 'I'm dying, Addie. Be lucky if I have six months.' He managed a sad chuckle. 'If you can call that luck.'

The silence that lay between them was the same silence that had remained unbridged for ten years. Addie gazed at him for a very long time before she leaned over to pick up the map from the bed and said, 'Tell me on the way up.'

CHAPTER EIGHTEEN

The woods lay silent under their thick blanket of snow. Large snowflakes drifted down through the trees as they crossed the stream and began the steep ascent towards the old military road somewhere far above.

Addie made no attempt to outpace him this time, and the only sound to disturb the still of the morning was the air they sucked in and breathed out, and the rush of white water somewhere nearby as it fell from Grey Mare's Waterfall to break over the jumble of rocks below.

They paused after a while to look back towards the village and the loch. The hills were lost in cloud that seemed to come down almost to the water's edge, where reflections of the sky were drowned in shades of grey. Brodie sat down on a rock to catch his breath. 'You know, I've been here nearly two days now, and I've hardly seen a soul.'

'The village is like a graveyard in winter,' Addie said. 'Not many more than five hundred live here year-round now. You'll see folk at church on a Sunday, or at the Co-op when you go for your messages. And if you head down to the pub at night, there's usually someone there you know. But when the weather's like this, people just tend to stay indoors. I guess it was different when the smelter was still on the go, and from all accounts it was like the gold rush when they were building the nuclear plant. Changed days, though.'

It was as if by speaking of things inconsequential, they might avoid addressing the elephant in the woods.

Brodie inclined his head to look up through the tall pines towards the sky. 'Not a breath of wind,' he said.

She nodded. 'The calm before the storm.'

He turned to look at her. 'Another one?'

'A biggie. Coming in off the Atlantic. They're forecasting hurricane-force winds. Rain turning to ice. And eventually to snow. We'll be buried in it here. And probably lose power again. Storm Idriss, they're calling it.'

'We'd better move, then.' Brodie got stiffly back to his feet, and they started once more over rough ground. Snow lay in patches, and winter-dead ferns bowed their heads under the weight of it.

Without looking at him, she said, 'So, are you going to tell me?'

He summoned courage and strength from diminishing reserves and after a few more steps, said, 'You know how your mother and I met, don't you?'

And he heard the hint of sarcasm in her retort. 'You rescued her from an abusive relationship.'

Brodie was unaccountably irritated. 'He was a drunk! And with a drink in him, he was violent.'

She muttered under her breath so that he barely caught it, 'Another addictive personality.'

'What?'

She shook her head and breathed exasperation like smoke into the cold. 'Nothing.' And quickly refocusing, said, 'So you got Mum's drunken partner put away.'

He stopped, taken aback. 'Is that what she told you?'

She shrugged. Neither confirmation nor denial.

He said, 'Lee Jardine was sent down for twenty years for drink driving.'

Addie glanced at him sideways. 'That seems a bit extreme.'

'He crashed his car into an SUV, killing a mother and her two children.'

Which stopped her in her tracks. 'Jesus,' she said, forgetting that she was supposed to be the sceptic here. Then she recovered herself. 'That must have been very convenient for you. With the competition out of the way, my mother was all yours.'

Brodie said, 'Think yourself lucky, Addie. Mel would never have left him. And she was already pregnant. So he could have ended up being your dad. And no doubt when he got drunk on the weekend, you'd have been on the receiving end of his fists, too. Or worse. Your life would have been very different.'

She stared at him, horrified by the thought, then was struck by another. Something unthinkable. 'I'm not . . .' She could hardly bring herself to give voice to it. 'I'm not his, am I?'

'Your mother swore not. And I've never had any reason to disbelieve her.' Even though the tiny seed of doubt somewhere deep inside him had never quite gone away.

But the thought clearly wouldn't leave her, and he could see all the uncertainty gathering like a storm behind her eyes. She turned abruptly and started off again through the trees, long legs powering up the incline so that he struggled to keep up with her. Then she stopped again, turning as he finally caught up. 'Why are you going to die?' she demanded.

He shrugged. 'Does it matter?'

She thought about it, then shook her head. 'Probably not.' She paused. 'Cancer, I suppose.'

'Isn't it always?'

She pursed her lips. 'So what happened between you and Mum? I want the truth. I deserve that.'

'You do, Addie. And it's all I've ever wanted to tell you.' He hesitated. 'But you won't like it.'

The storm bubbling up behind her eyes was as ominous as the clouds gathering overhead. 'Try me.'

He drew a deep breath, and steadied himself on the incline with his climbing stick. 'You were seventeen, Addie, when Jardine got out on licence. The moment I heard about it, I knew things would end badly. He'd had such a . . .' He searched for the right words. 'Such a Svengali-like hold over your mother . . .'

CHAPTER NINETEEN

2040

From the moment Tiny told me about Jardine's release, I had this sick feeling in the pit of my stomach. Just the thought that he was out there, walking the streets again, made my blood run cold. I couldn't get that old Scots aphorism out of my head: *ye're a lang time deid*. That woman he killed, and her two children. They were still dead. They would never walk the streets again. And here he was, out, with half a life still ahead of him.

I should have known better, but I did everything I could to try and keep the news of his release from Mel. I was scared how she might react. Even after all these years, and the life we'd made together, I was still afraid of the hold he had on her. The hold he might still have on her.

We moved in different social circles then. We lived in Pollokshields. Addie had gone to a good school. And she'd just started university. It was inconceivable to me that Mel would hear about Jardine's release by chance. Or that we would bump into him on some night out in town. We were regulars at the Theatre Royal now, and the bars and restaurants round about, which were not establishments that a guy like Jardine was likely to frequent.

I don't know why I didn't think of it, but of course I should have known that he would seek *her* out. And, I mean, she wouldn't be that hard to find.

I said nothing about Jardine, and life went on like before. At first I thought it was all going to be okay. But gradually, I became aware that Mel was changing. It was so subtle at first that I didn't notice. I can't even remember now how long it was before I did.

There was an increasing lethargy about her. She became tense and irritable. I saw that she was drinking more in the evenings, even when I wasn't there. She hardly ever laughed, and I had become so accustomed over the years to the peel of her laughter ringing out around the house. And, even then, it didn't occur to me why. I'm so bloody stupid! Maybe if I'd cottoned on sooner . . .

It was pure chance in the end that I stumbled on the truth. She'd left her phone on the kitchen table. I don't know where she was. Somewhere else in the house. But it chimed. You know, that sound it makes when there's an incoming text. The notification appeared briefly on her welcome screen, and I had time to read it before it vanished, lost behind a passcode that she'd always kept from me.

Leonardo, Friday at 7. L.

I didn't twig initially. She'd told me she was meeting a girl-friend for drinks on Friday. Leonardo? Had to be some kind of pub in town. Then it occurred to me that the girlfriend's name was Sarah. So who was L? And then it dawned on me with an awful clarity. Lee. It was Lee! He was meeting her at seven on Friday at somewhere called Leonardo. I went online and googled it. The only thing I could find was the Leonardo Inn. I knew the place. Out west on the Great Western Road. It had been there for donkeys. In the old days it had been known as the Pond Hotel.

I could scarcely believe it. Until I ran my mind back over the previous few weeks. The number of times she'd been meeting this girlfriend or that. And it felt then like my life had just ended.

I didn't even notice her coming back into the kitchen. Didn't hear her when she spoke to me. At least, not the first time. Then I heard her saying, 'Cam, Cam, are you with us?'

I looked up, and she had her phone in her hand. 'Sorry,' I said. 'I was away in a dwam.' But she was the one not listening now, as she read her text, and suddenly slipped her phone in the back pocket of her jeans like it was burning her fingers. I knew then that I had lost her.

She thought I was on duty Friday night. But I swapped shifts, and I was sitting in my car at the back of the parking lot at the Leonardo Inn for a good half hour before the seven o'clock rendezvous.

I saw Jardine arrive first in a beat-up old Tesla. So they'd given him his licence back, in spite of everything. Either that, or he was driving illegally. But long gone was the flash, expensive sports car. Which no doubt was something he could no longer afford.

He got out of his banger and propped his arse on the bonnet to light a cigarette and wait. She was nearly ten minutes late, finally arriving in one of the new generation of black e-cabs. Dressed to the nines and all made up for the big date. Lipstick, eye shadow, the lot. As if she needed somehow to impress him. Thirty-six years old, and she was still a fine-looking woman. But I loved her in her baggy old jog pants and T-shirt, without a trace of make-up.

I sat there behind the wheel of my car with tears filling my eyes. It was loss I felt more than anger. I could never be angry with Mel. She was so innocent. Even in her betrayal.

Jardine threw away his cigarette and when she ran to him, they kissed. Not just a casual, 'hi there' sort of kiss. It was longer than that, lips that lingered, turning the knife in me. As if I wasn't hurting enough. Then they laughed, and he held her hand as they ran lightly up the steps to push open glass doors into reception.

I sat for what must have been ten minutes or more, knuckles turning white as I gripped the wheel in front of me. What was I going to do? Turn around and drive away? Accept that life with Mel as I'd known it was over? It would have been impossible for me to pretend that I didn't know about her and Jardine. If I wanted to keep her, I was going to have to fight for her.

The girl at reception was flustered when I thrust my warrant card in her face and demanded to know the room number of the couple who had just checked in. She wasn't to know that I didn't have that authority. And she didn't need to check her records. She remembered it. Room 347.

I took the elevator up to the third floor, trying to not even think about what I was doing, holding every emotion in check. I was like some kind of container under pressure, ready to explode. I walked along a carpeted hallway and stopped in front of Room 347 to rap on the door. There was no magic eye in it, so he wouldn't be able to see me.

'What is it?' his voice barked from somewhere inside the room. A gravelly, smoker's voice grated raw by years of alcohol abuse. I wondered how he'd managed inside, but prison security was like Swiss cheese in those days.

I put on a posh voice. 'Richard from reception, sir. There's a problem with your electrics.'

'What the fuck?'

'The management's apologies, sir, but we'll have to move you to another room.'

I heard banging about behind the door before it flew open, and a semi-dressed Jardine filled the frame of it. There was no time even for surprise to register on his coupon before I put my shoulder in his chest, and we both went barrelling backwards into the room.

I heard Mel scream as I landed on top of him, and his foul breath exploded in my face. Just as all my pent-up fury exploded in the fists I slammed into his. I reckon I broke his nose and took out a couple of teeth with the first three blows. Then I punched him in the throat and he couldn't breathe. He was bucking beneath me like a demented horse, and I kept hitting him till I couldn't see his face for blood.

I was barely aware of Mel screaming at me, trying to pull me off, before finally the veil of madness lifted and I got to my feet with bruised and bleeding knuckles. Jardine lay on the floor gasping for breath, blood bubbling from between split lips.

I pulled myself free of Mel's grasp, and my eyes must have been on fire, because she recoiled from me as if I might hit her. As if I would. As if I ever would. Her blouse was open and I could see the black lacy bra against the white of her skin beneath it. Everything I didn't want to see. I grabbed her jacket from the bed and told her to put it on. She was coming with me.

It was only later, I guess, that I realised I had no right. That by removing her choice, I wasn't treating her any better than Jardine. I'm ashamed now of what I did. But when I look back on it, I'm not sure I would have done anything different. If only I could have spared Mel the hurt and humiliation.

She snatched her bag from the dresser as I dragged her from the room. Jardine had pulled himself up on to one elbow by now. And I could see only murder in the eyes that blinked away blood. 'I'll fucking get you, Brodie. Count on it. You'll fucking regret this, both of you.'

I pulled the door shut on him and hurried Mel away down the hall to the elevator.

The girl at reception stared at us, wide-eyed, as we ran across the lobby and out into the dying light of the day. I no longer had to drag Mel behind me. She came without resistance. That passive acceptance she always had of everything that life threw at her.

We sat for a long time in the car without saying a word. Staring sightlessly out of the windscreen, breathing hard, filling the air with our spent oxygen and all our regrets. When I finally turned to look at her, silent tears ran freely down her face. She said, in a voice so brittle it damn near broke my heart, 'I'm so sorry, Cam. He . . . he threatened Addie if I didn't see him.' And I thought there didn't seem anything threatening in the kiss I'd seen them exchange just fifteen minutes earlier. Maybe she read my mind, because she said, 'It doesn't mean anything.'

And I closed my eyes to shut out the pain, because I knew it meant everything.

'I can't even explain it . . .' Her words came staccato through her sobs. 'He . . . he just has this hold on me.'

I dropped my head on to bloody hands clutching the steering wheel. I whispered, 'Tell me you won't see him again.'

'I swear, Cammie, I swear it.'

But I knew she would.

CHAPTER TWENTY

2051

Addie climbed now in silence, her face as white as the snow in which she left traces, just a little colour rising in patches high on her cheeks. She had listened to her father in silence as they stood on the incline beneath the pines, before turning without comment to continue the climb towards the old military road somewhere up ahead.

Brodie felt hollowed out, as if letting go of everything he had kept to himself all this time had left a gaping hole inside him. Nothing rushed in to fill the void. Not even regret. And he wondered how something as full of nothing as emptiness could weigh so heavily.

Wearily he started off after her and they climbed in silence for another fifteen minutes or more, before emerging finally from the trees and on to the unbroken snow-covered military road that cut its way around the side of the hill. It was exposed here, and they could see all the way back down into the valley. The mountains that rose steeply from the banks of the loch pierced a troubled sky, and the clouds which had earlier obscured them seemed anxious now to pass them by, blown east on the first breath of the coming storm. Brodie felt it, too, in his face. Like an icy hand brushing cold flesh.

'Not far now.'

He heard Addie's voice, and turned to see her striding off along the road towards the corrie they had passed through yesterday on their way down from Binnein Mòr. He hadn't known what to expect from her. Some reaction, at least. Not just silence. It was as if his words had run off her like water on wax. He had no idea whether she was in denial, or simply processing. But her lack of response left him feeling, if anything, emptier than before.

They walked for a good ten minutes in further silence, Brodie trailing twenty yards behind. Until they reached an area of unbroken snow that lay to their right. A turning or passing area, perhaps, on this single-track road. She stopped and waited for him to catch up.

'This would be the last place he could leave his car without blocking the road,' she said.

Looking ahead, Brodie could see the road meandering up the hill towards a hairpin that turned across the Allt Coire na Bà, or one of the streams that fed it. Beyond this open area on the right, the tree-covered hillside fell away steeply, and they could hear the sound of running water from far below. Wet snow creaked under his feet as he crossed towards the drop, and a couple of startled ptarmigan with their pure white mass of winter plumage clattered noisily away into the forest.

Addie joined him on the edge as he gazed down into the trees that sparsely covered the top part of the slope. 'If he had left it here, they'd have found it very quickly,' she said.

He held out a hand towards her. 'Help me down.'

She took it almost without thinking, and braced as he stepped cautiously down on to the slope, to drop her hand and wrap an arm around a gnarled pine to stop himself slithering off towards the stream.

'What are you doing?' she called after him as he slid then from one tree trunk to the next.

'Looking for Younger's car,' he called back.

He stopped, crouching down beside the nearest trunk, and ran a hand lightly over the bark near where the roots had spread themselves out to gain a foothold on the hillside. He turned his face back up the slope towards her.

'White paint,' he shouted, and heard his voice echo around the ravine below. He stood up, supporting himself against the trunk, and peered off into the gloom. 'Looks like someone might have driven it, or pushed it off down here.' He sat down abruptly in the snow, and braced and bent his legs in turn to lower himself down the incline on his backside.

The noise of running water grew louder as he neared the foot of the drop, and he saw the hulk of a white vehicle, half-buried in snow. Its nose was sunk into the bed of the stream, breaking the flow of water and sending it in white spate around either side of it.

He turned, hearing Addie arrive behind him. She had negotiated the slope much more quickly than he, and stood breathing hard, supporting herself against the nearest trunk. She gazed in wonder at Younger's car, lying as it was at a crazy angle, its rear wheels and axle completely clear of the ground and backfilled by the snow drifting up around it.

Brodie said, 'No one was ever likely to find it down here, even in August. And now it's perfectly camouflaged by the snow.' He moved down to place one foot in the stream. 'Give me a hand to clear the snow away and we'll see if we can get into it.'

They worked slowly, careful not to dislodge the vehicle from its final resting place in case it fell on them. Finally, Brodie was satisfied that the door would open clear of the snow. He unzipped the breast pocket of his North Face and took out the keycard he had chipped free from the ice on the mountain.

Addie said, 'Surely there won't be any charge left in the battery?'

'Enough, hopefully, to read the card,' he said. 'If not, we'll have to smash the window.' He laid the card against the sensor in the column between the front and back doors, and heard an audible click above the rush of water. He tried the handle and the door swung open, suddenly, almost knocking him off his feet. He recovered himself quickly to scramble away in case the car dislodged itself. But it didn't move. He breathed a sigh of relief. 'It must be wedged solid.'

He pulled himself up with one hand on the roof, and swung himself into the driver's seat, tipped forward against the steering wheel. Addie slithered down to peer in beside him. It was ice-cold and dark inside. But there was enough light to see that a jacket tossed carelessly into the back seat had fallen on to the floor. The mats in the front were littered with chewing gum wrappers. A green scent diffuser in the shape of a Christmas tree dangled at an odd angle from the rear-view mirror, but it had long since lost its perfume. Strangely, there was a very human smell in the car. The faint fragrance of body odour and aftershave. The last traces that Charles Younger had left on this earth.

Brodie tried the glovebox, but it was locked electronically, and he thought there probably wasn't enough charge left in the car to boot up its computer to open it.

'Here.' Addie handed him a large hunting knife that she took from her daypack. The look he gave her brought a smile to her face. 'I should have been in the Boy Scouts,' she said.

But he couldn't find a smile in return. He took her knife, unsheathed it, and forced open the glove compartment with a splintering of moulded fibre. Inside were maps, and some notebooks with pages of scribbled shorthand. Brodie flipped through a few, but the strange markings made no sense to him. Then, beneath the car's leather-bound instruction manual, he found something that looked like the kind of mobile phone that people had used around the turn of the century. Chunky, yellow, and with a grey liquid-crystal display screen. He took off his gloves to examine it, turning it this way and that. 'What the hell?' He turned towards his daughter. 'Any idea what it is?'

She nodded. 'It's a Geiger counter.'

He frowned. 'For measuring radioactivity?'

'Yes.' She paused. 'Dad, what would he want with a Geiger counter?'

The fact that once again she had called him *Dad* without thinking stilled his heart and he couldn't meet her eye. But his only response was to shrug. 'No idea, Addie.'

He unvelcroed his parka to access an inside pocket, and zipped the Geiger counter and the notebooks safely away, then heaved himself out.

'Let's try the boot.'

It took some minutes to force the lock, and when finally they lifted the lid, they found only a spare wheel and a bag of tools.

Brodie said, 'There's probably storage space under the bonnet. But we'll not get access to that until we can get this thing winched out of here.' He sat down in the snow and rubbed his face, breathing frustration into his hands.

Addie squatted beside him. 'So what does any of this tell you?'

'It tells me that whoever killed him up on the mountain came down to get rid of his car before anyone spotted it.'

'But how? I mean, he didn't have a key, because that was still up there in the ice. And if the car was in park, then he couldn't have pushed it over.'

Brodie stood up suddenly. 'I wonder if Younger left it in sentry mode.'

'What's that?'

'Some cars have got a security system, Addie, that uses the self-drive cameras. There's usually about eight of them around the vehicle. If you leave them on sentry mode, they'll record anyone or anything that moves around it.'

'And there'll still be a record of that?'

'Let's find out.' He swung himself back into the car and leaned across the passenger seat so that he could reach into the back of the glovebox. Using his fingers as eyes, he felt around until they settled on a raised area at the left rear corner. With finger and thumb, he grasped the hard edge and tugged it free. He brought out his hand and held the object he had removed up to the light.

'What is it?' Addie peered through the gloom.

'An SD card. If Younger's car had sentry mode, and it was activated, whoever shoved it over the edge should be caught on video. And it'll be on this card.' It was a long shot, and he wouldn't know if there was anything on it until he got back to the hotel to slot it into his laptop, but it was time he had a break. Nothing else had gone to plan so far. He secured the card in another pocket and said, 'Let's get out of here.'

It was harder getting out of the gully than it had been getting in, and it was nearly ten minutes before they were standing in the parking area off the old military road, breathing heavily and perspiring in the cold air. The wind was getting up now, and Brodie felt it filling his mouth as he fought to recover his breath.

'We'd better get back to the village,' he said, and they started off back along the road until reaching the point where they had climbed up to it from below. Overhead, the clouds had morphed from ominous to threatening, and you could smell the coming storm on the leading edge of the wind.

It wasn't until they had climbed down through the trees to where the ground levelled off and the going got easier, that Addie formed words to express the thoughts that had been eating away at her all this time.

'So it was Mum who had the affair. Not you.' She wasn't asking, so he assumed that she had been processing it and was voicing it now as a statement of fact.

'Yes.'

'And that's what you wanted to tell me? That's what was so important that you deceived your bosses to get yourself sent up here?'

Brodie drew a deep breath. 'It's important enough, Addie. But it's still not the whole story.'

She looked at him. 'I didn't think it could be. Mum didn't kill herself just because she'd had an affair, did she?'

He shook his head. 'No.'

'So, are you going to tell me?'

'I will, Addie.' He hesitated. 'But there are things I need to do first. I need more time with you than we have right now.'

'Time for what?'

'To explain.'

'How you drove Mum to suicide, you mean?'

He glanced at her, expecting to see the hatred she'd harboured for him all this time still reflected in her face. But her expression was blank. Eyes cold, emotionless, and assiduously avoiding his.

'Yes,' he said.

They stood there for a long time listening to the wind, and Brodie thought how his story was just like that wind. Cold and unforgiving, and gone in the blink of an eye. Like his life. They walked then the rest of the way in silence until they reached the Grey Mare's car park. And stopped at the parting of the ways.

'So,' she said. 'What now?'

'I need to get to my laptop to see what, if anything, there is on this card. And if the phone or internet is back up, then I need to check in with HQ in Glasgow. We need a team up here. There's two people dead and a killer still on the loose.' He closed his eyes as he felt the pressure of it all weighing down on him. 'I'll have to come back to the police station sometime this afternoon. I need to take a look at that CCTV footage of Younger and the unidentified individual he was talking to in the village the day he disappeared.' He paused. 'Maybe we could talk then.'

'I'm not sure I want to hear what it is you have to say. Whatever it is, maybe it's better if it dies with you.'

She turned abruptly and walked away in the direction of the police station.

CHAPTER TWENTY-ONE

As he climbed the slope from the football field to the hotel, he could smell woodsmoke carried on the wind, and saw curls of blue smoke whipped into the gathering storm from the chimney top above the bar. Brannan's SUV was parked at the foot of the steps. For once, Brodie thought, there was someone home.

He kicked the snow from his boots on the top step and pushed open the door into the entrance hall. Brannan emerged from the bar. He must have been watching Brodie's approach unseen from behind reflections on glass.

His smile was forced. 'Internet's back online. Mobile phones, too.'

'Good,' Brodie said.

But Brannan made a face. 'We're not likely to have them for long, though. Storm Idriss is scheduled to hit in a couple of hours, and it'll probably take everything out again.' He flicked his head back over his shoulder. 'I was just trying to build up some heat in the bar. In case we lose power again, too.' He laughed at his own optimism. 'In case? I should say "when".'

Brodie said, 'You've spoken to Jackson?'

Brannan's face clouded. 'I haven't had a chance.'

Brodie's eyes turned dangerous. 'Oh, yeah, cos you're so busy here at the hotel.'

Brannan said quickly, 'No, what I mean is, I haven't been able to reach him, Mr Brodie. He's at the plant. Won't come off shift till six. There were no phones all morning. And it's hard to get a call through to him there anyway.'

'Then try harder.'

'I will, I will . . . But, you know, I promised Joe confidentiality.'

Brodie took a step towards him. 'If there's no rendezvous arranged by close of play this afternoon, I'm going to arrest you for obstruction of justice, Brannan.' He scrutinised the man's frightened face. 'Do you understand me?'

Brannan nodded.

'Good.' Brodie started for the stairs, then stopped and turned. 'One other thing.'

Brannan eyed him warily.

'What does he look like, this Joe Jackson?'

Brannan frowned. 'I don't—'

Brodie cut him off. 'Just describe him to me.'

Brannan looked perplexed, then almost pained as he tried to pull an image to mind of the man he had spent half the day with just yesterday. The succession of witnesses over the years who had struggled to recall the details of events which had unfolded in front of their eyes meant that Brodie was no longer surprised by people's faulty memories. 'He . . . he's tall. Probably six foot. Maybe a bit more.' He raised a hand to his own balding head. 'Losing his hair. Sort of gingery, going white.' He was warming to his memory. 'A wiry guy, not much meat on the bones.'

Brodie nodded. This was better than he had expected. 'Talk to him,' he said, and turned to run up the stairs.

In his room, Brodie unfolded his laptop on the dresser and booted it up. He took the SD card from his North Face and examined it in the light. Extended capacity. Ten terabytes. Enough for hours of 6K video. He slipped it into the card slot on the side of the laptop and opened it up on-screen.

Younger's housekeeping had been poor. There were hours of recorded video that he hadn't bothered to wipe. Fortunately there was a date stamp, so Brodie was able to fast-forward to the day the journalist went missing. It was 9 p.m. when the cameras on Younger's car flickered into life and Brodie saw a figure approaching from the rear. A man wearing a hoodie and jeans, and to Brodie's disappointment, a ski mask – aware of the possibility that he was being captured on camera. As he moved around the car, his image segued from one camera to the next. He tried each of the doors, but there was no way of getting into the car without breaking a window.

A hand came into close-up as the man turned away from the driver's door, and Brodie stiffened. He froze the image and zoomed in on it. It was some kind of work glove. *M-Pact*

Mechanix. Brown and tan, reinforced across the knuckles and along the back of each finger. With four distinctive horizontal slashes at each joint to allow for easy flexing. The same pattern that, with repeated blows to Younger's head, had been imprinted in clear contusions in the flesh of his face.

Brodie switched applications and googled *M-Pact.* He found the glove in seconds. *Impact Guard™ for shock protection.* And *TrekDry® material to keep hands dry.* The reinforcement was provided by thermoplastic rubber, and something called EVA foam protected all the joints. Ideal for heavy mechanical work. Or mountaineering.

He switched back to the video and set it to play. The wearer of the gloves simply walked away, moving quickly out of shot. The cameras recorded for another thirty seconds, then stopped, and the image went black. There had to be more. Brodie waited.

When the recording restarted, an hour had passed, and the light had faded. The picture was grainy now. The movement which had triggered the recording was the arrival of another vehicle, which immediately doused its headlights. It swung quickly into position directly behind Younger's car, and it was impossible to tell whether this was an SUV or a pickup. Even the colour of it was difficult to determine. Dark blue or green. Maybe even grey. Its driver had taken the precaution of covering the licence plate, but there were bull bars on the front.

Younger's car juddered as the vehicle behind it engaged,

then began inching it forward. The wheels would have been locked, but the superior power of the other vehicle easily pushed them across the gravel.

Suddenly the view from the rear cameras angled towards the sky, and all the images recording on to the SD card became blurred as the vehicle tipped down the slope, gathering momentum, and jarring as it struck several trees on the way down. It seemed to take an eternity to reach the bottom of the drop, but in fact was recorded as being fewer than five seconds. The downward progress of the car ended suddenly as the nose buried itself in the stream. The front cameras registered underwater pebbles worn smooth by eons as the cloudiness of the impact quickly washed away in the flow of the stream. The view back to the top of the slope revealed the distant silhouette of a man standing against the light of the stars. He waited for only a moment before turning away out of shot, and less than half a minute later, the recording stopped and the picture went black. There was no further video on the card.

Brodie was startled by a knock at the door. 'Yes?'

It opened, and a hesitant Brannan took a couple of steps into the room. 'Sorry to disturb you, Mr Brodie.' And Brodie realised for the first time how quickly the light was fading. The cloud was almost black beyond his window, the afternoon light sulphurous, the wind rattling the window frame.

'You spoke to Jackson?'

Brannan nodded. 'He's agreed to meet you on the proviso that you'll keep his name out of it.'

'There's no way I can guarantee that, Brannan.'

Brannan made a face. 'I thought that. But I'll let you tell him.'

'Is he coming here?'

Brannan shook his head vigorously. 'No. He'll meet you tonight. 8 p.m. On the north side of the loch. A couple of miles short of the power plant. Down off the road there's a concrete bunker which provides an emergency escape from the storage tunnels below. I can show you on the map.'

Brodie sighed. 'I'd rather he just came here.'

'He won't do that, Mr Brodie. Whatever it is he knows, whatever he told Younger, he's scared. I mean, really scared.'

'And how am I supposed to get there?'

'I'll lend you my SUV.'

'You could just drive me.'

Brannan shook his head. 'I have a large party booking in for Christmas this year, Mr Brodie. The organiser is calling tonight with details. I need to be here to take the call.' He added quickly, 'It's only about a ten- or fifteen-minute drive.'

When Brannan had gone, Brodie sat in the gloom of his bedroom for several minutes, turning everything over in his mind. The autopsy, the gloves, the car in the ravine. Sita's murder. It pained him every time he recalled the image of the pathologist folded into the cake chiller. Something she had found during the post-mortem had made her a target for the killer. Something that might reveal his identity, or the motive for Younger's murder. But whatever it was she'd found, it was lost now. All her samples vanished, along with Younger's body.

He checked that he still had internet and slipped on his iCom glasses. He would record his report and send it to Pacific Quay so that he didn't have to engage. That way he could register every detail without interruption, right down to and including his rendezvous later with Joe Jackson. If Storm Idriss brought down power lines again, as Brannan predicted, it might be the last chance he had to report in before sometime tomorrow, or later.

CHAPTER TWENTY-TWO

Lights blazed in most of the windows of the police house, shining into the gloom of the dwindling afternoon. Brodie had not seen a soul on his trudge round to the village from the hotel, apart from the occasional vehicle passing on the road. Where cars had travelled and people had walked, wet snow had turned to ice in the plummeting temperatures, and was treacherous underfoot.

He pushed open the gate and walked up to the door of the annexe with a sense of trepidation. If Addie was determined not to listen, then how could he tell her anything? And certainly not in the presence of her husband or her son. He closed his eyes and drew a deep breath. There were other things that took precedence. He needed to focus.

He opened the door into the warmth of the little police office and saw that it was empty. The computer screen was illuminated by a screen saver, and an anglepoise lay a circle of light on the desk. He closed the door behind him, shutting out the howl of the wind, and immediately heard raised voices coming from the house.

A man's voice, which must have been Robbie's. And Addie. Shrill and accusatory. He couldn't make out what they were saying, or anything that might provide a clue to the cause of the argument. The plaintive wailing of young Cameron, distressed by the raising of his parents' voices, made it difficult to hear them clearly.

He stood for a little while, wondering what to do, before opening the door to the house and calling, 'Hello?' The voices of the adults immediately subsided, but Cameron's wailing provided the continuity of the argument in its aftermath. After a furious exchange of stage whispers, Robbie emerged into the darkness of the hall from a lit room behind him, and hurried through to the police station. He was flushed with embarrassment as he came in, quickly closing the door behind him. And the sound of Addie comforting her son was reduced to a distant murmur.

'Sorry, Mr Brodie,' Robbie said, attempting a smile that didn't quite come off. 'Domestic bliss.' And he could only have been too acutely aware of the irony in his addressing this to the father of the woman he'd been fighting with.

Brodie said, 'Tell me about it.' Then, 'I want to have a look at that CCTV footage.'

'Oh, yeah, of course.' Robbie rounded the counter and sat down at the computer. A swipe of his mouse banished the screen saver, and he accessed the hard drive to search for the archive footage.

Brodie came around to stand behind him. He said, 'I have

a meeting tonight with a guy from Ballachulish A who seems to have been some kind of contact of Younger's.'

Robbie swivelled round in his seat. 'A contact?'

Brodie shrugged. 'Brannan seems to think the guy might have been a whistle-blower of some sort.'

Robbie looked nonplussed. 'Blowing the whistle on what?'

'Hopefully that's what I'll find out tonight.' He paused. 'But it might explain why Younger had a Geiger counter in his car.'

'Yeah, Addie told me you'd found the car. How the hell did it get down there?'

'Someone shunted it over the edge. The whole thing was captured on video by the car's sentry mode.'

Robbie raised hopeful eyebrows. 'So you saw who did it, then?'

Brodie shook his head. 'He was wearing a mask when he first examined the vehicle. He came back with an SUV or a pickup truck to push it down into the gully, but by then it was too dark to really see much.'

'Fuck!' Robbie said. And was immediately self-conscious. 'Sorry, sir.'

Brodie managed half a smile. 'I think I've heard grown men swear before, Robbie.' He nodded towards the screen. 'What have we got here?'

Robbie turned back to the computer. 'It's only about thirty seconds or so.' He brought the video up on-screen and hit the play arrow. The image flickered, then cleared to reveal Younger in full 6K detail standing in the street, somewhere near the

Co-op, talking to another man. It was a bright summer's day and the light was good. The conversation was animated, and ended with a laugh and a wave before each man exited the frame in different directions.

'And you've no idea who the other man is?'

Robbie shook his head. 'Never set eyes on him before.'

One thing was clear to Brodie: if Brannan's description was to be trusted, it wasn't Jackson. This man was shorter than Younger and had a full head of light brown hair. Younger himself looked almost grey, his complexion pasty and pale. And although he joined the other man in perfunctory laughter, it was clearly forced. He looked to Brodie like a man with a lot on his mind.

It occurred to him that they might send the video to a lip-reader to learn what they had been talking about, but that thought was quickly superseded by another. He said, 'Play it again when I tell you,' and slipped on his iCom glasses.

Robbie looked at him curiously. 'What's this, new tech?'

Brodie nodded. 'Play it,' he said. Then, 'iCom, scan the video.' And he focused on the computer screen, ignoring the green heads-up display that his glasses scrolled across his vision. Until thirty seconds later, when it red-flagged the video as fake.

This time Robbie watched him, rather than the screen. 'What does it tell you?'

'That this video is not genuine.'

Robbie scowled. 'Well, I don't see how that's possible. It's what was recorded by the CCTV camera.'

Brodie ignored him and said, 'iCom, connect to the local server and upload the video.'

iCom told him in his earbuds that it was searching, then uploading. It took less than a minute.

This time Brodie instructed it to strip back the AI neural generator to reveal the original scan carried out by the discriminator. The process was almost instantaneous, and the video reran itself, visible only to Brodie in his glasses. 'Jesus!' The oath escaped his lips in a whisper.

'What is it?' Robbie was searching the reflections in Brodie's glasses as if he might be able to catch sight of the video in them.

'I'll show you.' And Brodie instructed his iCom to download the stripped-back video to the computer.

The file appeared on Robbie's screen and he clicked to play it. The video seemed unchanged, except that Younger was no longer talking to a man with a full head of light brown hair. He was in animated conversation with the man who that morning had repaired the charging cable of the eVTOL on the football field. Calum McLeish.

Robbie sat in stunned silence. His whispered oath was barely audible above the sound of the wind outside. 'Fuck!' He turned towards Brodie. 'I don't understand, sir. How is that even possible?'

Brodie said, 'Advanced GAN software. Makes it easy to substitute one face for another. It's probably not even a real person. More likely an artificially generated face.

Robbie was bewildered. 'But who the hell could have done that?'

Brodie said, 'Did McLeish ever have access to this video?'

'No.' Then he paused. 'Although he does have access to the system. He has a contract with Police Scotland to service the CCTV in the village. Cameras and computer. So I suppose he could have.'

Brodie was remembering the dark blue pickup truck that McLeish was driving when he came to the hotel. He said, 'Tell me where he lives. I think it's time Mr McLeish and I had a wee conversation.'

Brodie had covered about two hundred yards, heading south on Riverside Road, parka zipped up to the neck, hood yanked on over his baseball cap. He could hear nothing for the roar of the wind in his ears, and was startled when Addie fell into step beside him. He stopped in his tracks and turned to take in the troubled set of her pale face. She had obviously pulled on her ski jacket in a hurry, and hadn't even zipped it up. Her hair blew wildly around her head in the wind.

'I'm sorry you had to hear that,' she said, raising her voice to make it heard above the wind. 'I don't know how much of it you actually made out, but—'

'Darling, I heard nothing but raised voices. Husbands and wives fight. It sounded pretty heated, but I've no idea what you were arguing about.'

She seemed almost relieved.

'But if I were to guess,' he added, 'I'd think it might have had something to do with what you said earlier today.'

She frowned. 'What? What did I say?'

'Something about another addictive personality?'

Her mouth gaped slightly. 'If you didn't hear what we were arguing about, how could you possibly know that?'

He sighed. 'Many years as a student of the human condition.' He hesitated. 'It's not drink, is it?' The idea of history repeating itself in that way would have struck just too close to home.

She shook her head and averted her eyes. 'He's a gambler.'

Brodie's heart sank. He'd seen only too often over the years how a gambling addiction could destroy a life, wreck a marriage. He took her by the shoulders. 'He doesn't . . . he's not violent, is he?'

'No. No, never. Robbie's not like that.'

Brodie sighed his relief into the wind. That would have been one irony too many.

'He's just . . . well, hopelessly addicted. He's put almost everything in hock to feed his habit. Online. Always online. It's so fucking easy. I think they rig it. A small win here, a small win there, just to keep you at it. Then they suck you dry.' She paused to catch her breath. 'It's been a nightmare for me and Robbie. He's totally out of hand. We're overdue on most of our bills. If the house didn't come with the job, he'd have mortgaged that, too, just so he could feed the habit.'

'You know there's counselling available.'

She shook her head despondently. 'You can't get help until

you admit there's a problem. I've tried, Dad, believe me. But he won't listen. Refuses to accept that he has an addiction. At least, not to me. But he must know it in himself. I think he's desperate.'

She almost staggered as the wind gusted upriver from the loch, and he drew her into the lee of a building.

And now it all came pouring out of her. Everything that she must have been bottling up for weeks and months. With no one to tell. Too humiliating, perhaps, to admit to friends. Maybe even more humiliating to admit to the father she hadn't spoken to in ten years. And yet, here she was baring her soul. Seeing him, maybe, as the last and only hope of salvation.

She said, 'I've done a lot of online research, looking for answers. But it all seems pretty hopeless. Ever since the British government legalised online betting with their Gambling Act in 2005, it's just got out of control. Yeah, sure, it's raised billions in taxes over the years, and earned billions more for the gambling industry itself. But it's created millions of addicts. Annual suicide rates run to hundreds. That's thousands of people who've killed themselves since then, just so the treasury can raise money in easy taxation. Legislation introduced by self-professed Christians.' She paused and almost spat her contempt into the wind. 'Fuckers! And what's the phrase they use in all that TV advertising? *Remember to gamble responsibly*. Fuck! Dad! That's like telling an alcoholic to drink responsibly. Fucking, fucking hypocrites.' And the tears came bubbling out of her eyes as he drew her into his arms and held her close. He could feel the sobs juddering through her body, and he remembered holding a sobbing Mel in his arms, too.

They stood for a long time holding each other, buffeted by the wind, her tears soaking into his North Face, until finally she drew away and looked at him with desperation in her eyes. 'It's not really Robbie's fault. He's a victim. It's an illness.' And then came the hesitation. But it didn't last long. 'Will you speak to him?'

Brodie felt himself almost physically withdrawing. He'd have done anything to help her. But a third party intervening between husband and wife never, in his experience, turned out well. 'It's not my place, darling,' he said. And saw her face harden.

'You're a senior officer.'

'I have no jurisdiction over Robbie.'

'Then as the father of his wife.' And that came like a blow to his solar plexus.

'Addie . . .' He didn't have to give voice to his doubts. It was in his eyes, his whole body language.

She took a step away, gazing at him with all the hatred he remembered from way back. Hostility filled her eyes along with the tears and humiliation. She didn't even wait for him to reason with her. 'Well, fuck you, then.'

She turned and strode off, back along Riverside Road, her hair and her open jacket billowing out behind her. He had let her down. Again. He sighed deeply and screwed up his eyes, knowing, too, that any intervention by him, professionally or personally, was not going to fix the problem. Would almost certainly make it worse. And yet hers had been a heartfelt cry for help. How could he refuse her? He opened his eyes, lifting them to the heavens, and knew that somehow he was going to have to speak to Robbie.

CHAPTER TWENTY-THREE

McLeish's house backed on to the community fire station, a bungalow with a double garage attached. The footpath from the gate had been meticulously cleared of snow, and Brodie crunched over granite chippings to the front door. To the right of it, a light shone out from the living room window into the gathering darkness of the late afternoon. He was about to knock when he noticed that there was also light spilling from the open doors of the double garage off to the left, and he walked around the front of the house to look inside.

An old petrol-engined Porsche, a classic car from the 1970s or eighties, lay in pieces on the floor, the body of it jacked up for access underneath. A scarred wooden bench that ran along the back of the garage was strewn with tools and cans of oil and dirty rags. The wall behind it was hung with power tools and cables and saws. To the left of the Porsche lay the vacant space where McLeish clearly parked his pickup, a charging point and cable fixed to the wall next to it. It was conspicuous by its absence.

Brodie stepped over an open toolbox, careful not to stand

on contents which were scattered across the floor – spanners, screwdrivers, wrenches. He walked to the bench at the rear of the garage. His eye had been drawn by the brown and tan of a pair of well-worn work gloves lying next to the vice. The fingers of each glove were curled in towards the palm, almost as if there were still hands in them trying to grasp something unseen. He lifted up the right-hand glove and read the maker's name on the back of it. *M-Pact Mechanix*. And there were the four slashes in the finger reinforcements at each knuckle joint to allow for flexing. Forming the same pattern that Sita had found imprinted on Younger's face by his attacker.

'Can I help you?' The woman's voice was sharp but wary, and startled him.

He turned to find a middle-aged woman in jeans and sweatshirt standing in the frame of the open garage door. Once dark hair was streaked with grey and drawn back into a knot behind her head. He put her somewhere in her middle fifties.

He lay the glove back on the bench. 'I was looking for Calum McLeish.'

She eyed him suspiciously. 'And what exactly is it you would be wanting with him?'

Brodie stepped towards her. 'I'm sorry. Mrs McLeish, is it?'

'It is.'

He fumbled for his warrant card in an inside pocket and held it towards her. 'Detective Inspector Brodie. Your husband helped me out with the repair of a charging cable earlier today.'

She seemed relieved. 'Oh. Yes. He told me. He's gone up

to the hydro plant, Mr Brodie. He's got some more tools in his workshop up there. Something he needs to put this mess back together.' She waved a hand towards the deconstructed Porsche. 'He wanted to get there and back again before the storm broke.'

Brodie walked back up to Lochaber Road and crossed the bridge over the Leven to the south side of the village. Most of the services were on this side of the river. The post office, the Co-op, the boat club, several hotels and guest houses, the local housing authority. But nobody was venturing out into the coming storm. Brodie reckoned most folk would be cooried down at home, bracing themselves for Idriss. It was only fools like him who were out and about at a time like this.

He turned left on to one of the many old military roads that circled the village, past the National Ice Climbing Centre in the remains of the former smelting plant, a brewery, the Salvation Army church. On the bridge over the tailrace, he stopped and gazed up the three-hundred-metre length of it as it curved away towards the hydro plant. The rush of spent water was almost deafening as it made its way along this narrow canal from the turbines it had turned to generate power. Behind him the water turned white as it spewed into the River Leven. A good three metres beneath the bridge, it passed in spate, black-streaked and unforgiving, high stone walls rising on either side. Water that had been carried by gravity through pipes all the way down from the Blackwater dam in the hills above.

Brodie followed the tyre tracks in the road that ran alongside the tailrace, past an old stone-built hostel and a row of multicoloured lodges. Beyond the fence that topped the walls of the tailrace itself, unbroken snow lay across a large tract of land where the bulk of the aluminium factory once stood. On the hill above it, a dilapidated building of square white cubes, which at one time housed workers from the smelter, nestled in stark abandonment among the winter-bare trees. A brief incarnation as a military training centre had been cut short by the Scottish Government after independence. Empty and decaying, it stood now as a monument to a golden industrial age long since passed into history.

The hydro plant stood proud on the rise above the tailrace, tall windows in a long, narrow stone building rising up the gable end to the pitch of its slate roof. A large, bright blue roller door to the left of the windows allowed access for heavy machinery. It was shut. But a small door beneath the windows stood ajar, and light fell out from the glass above it to lie in elongated squares across the snow.

McLeish's dark blue pickup was parked outside. Brodie walked carefully around the vehicle to see what he hadn't noticed earlier in the day: the black-painted bull bars at the front of it. He crouched down, and in the light of the windows, saw that they were scraped and scuffed, with tiny streaks of white paint still ingrained in the front edges of the top and bottom bars.

He stood up with a grim sense of foreboding. Here was a man, he had no doubt now, who had killed twice. He had everything to gain and nothing to lose from killing again. Brodie approached the open door with caution. This was an emergency door that opened out, a push bar on the inside of it. He pulled it fully open and stepped into the plant. It stretched off into darkness, where three ten-megawatt generators powered by water from the dam produced as much noise as electricity. At the near, lit end, a pickup truck and Land Rover were parked on maroon tiles, and a green-painted walkway led past them towards a row of the original one- and two-megawatt generators, which had been preserved for posterity. Overhead, a large yellow crane that ran along steel beams set high on the walls hung silent.

To the right of the vehicles, lights shone in a Portakabin that Brodie assumed was an office or workshop. He leaned in through the open door. 'Mr McLeish?' His voice was greeted with silence. There was nobody here. He stepped out again on to the tiles and peered up into the darkness at the far end. McLeish had to be somewhere up there.

Now he raised his voice above the roar of the generators to call into the dark, 'Mr McLeish!' But even if he was there, Brodie realised, he wouldn't hear him over the thunder of the machines, and Brodie began to make his way carefully along the length of the building. The old, redundant Pelton turbines seemed almost to mock him with their silence. Ahead, great blue pipes like giant worms emerged from the working generators to burrow down below the building and feed water away to the tailrace.

He had almost reached the far end when the lights went out. Brodie froze where he stood, enveloped by sudden and absolute darkness, before what little daylight remained outside seeped in through the rows of skylights overhead to bring dark form to the shapes around him. He spun around, sensing that there was someone there, someone he wouldn't hear above the racket of the generators. But there was no one. Just the ghost of his own insecurity, insubstantial and lost in darkness. He started back towards the door, moving as quickly as he dared in the dark.

He felt more than saw the shadow of a man emerge from among the disused turbines, and turned to raise his arm just in time to stop a monkey wrench from splitting his skull. He felt pain, like red-hot needles, shoot up his left forearm and staggered backwards, crashing against some immutable piece of machinery that jarred through his entire body. His attacker came at him again, the monkey wrench raising sparks from the stone wall behind his head as he ducked to one side and it missed him by a hair. More in hope than expectation, he swung a fist into darkness and felt it connect with flesh and bone. He heard the man's grunt of pain, and capitalised on the moment to lunge forward, his shoulder connecting with his attacker's chest. The momentum carried them both backwards until they lost their footing and crashed to the floor.

Brodie heard the wrench clattering away across the tiles and went for the other man's eyes, but found instead only the smooth merino wool of a ski mask. A knee in his diaphragm took all his breath away, and he rolled over, gasping and choking back the bile rising in his throat. He heard the other man scrambling away across the tiles in search of his wrench, and with a huge effort of will, Brodie got to his feet and started running. Back the way he had come, towards the open door.

But after just a few paces, he could hear his attacker right behind him, breath rasping above the rumble of the turbines. There was no way he could outrun him, and as he staggered through the door into the cold outside air, he turned to face him. For a moment, in the dying light of the day, he saw murder in the other man's eyes. And this time it was his attacker who had the momentum. His shoulder powered into Brodie's chest, and both men fell backwards, locked in mortal embrace.

They crashed hard against the fence, tipping sideways over it, to fall together between iron cross-beams into the thrashing waters of the tailrace as it powered its way out of the building. The cold hit him like a physical blow, and both he and his attacker immediately released their grip on each other.

Now it was the water that held him and had all the momentum. Brodie was powerless to resist it, smashed from side to side against one stone wall then the other, swallowing huge quantities of water, choking and gasping for air. The speed with which it carried him away towards the river was relentless. His instinct, as it had been when caught in the avalanche, was to try to swim, even though the feeble thrashing of his arms and legs was worse than useless against the powerful currents of the tailrace.

His forehead struck the wall, and his head filled with light. He had lost his man, and knew he was losing his fight against the water. But this was no way for his life to end, with so much left undone, so much left unsaid. And yet the attraction of just closing his eyes and letting the cold and the water carry him off was almost irresistible.

He saw the bridge where he had stood only minutes earlier flash by overhead, and now the water turned white as its path broadened through a drop in the tree-lined riverbank and swept him into the swollen, snow-melt turbulence of the River Leven as it surged towards the head of the loch.

Suddenly he felt the depth of the water beneath him, and the unforgiving nature of its power as it swept him irresistibly towards his death. Yet still he fought for life, without understanding why, thrashing through the water as if his ebbing strength was in any way a match for it. He was numb now. All pain vanquished. He felt swollen and weighed down by his clothes, and completely at the mercy of the currents and eddies that tossed him freely this way and that. Now the water sucked him under, and for a brief moment, he believed he had drawn his last breath, the angry roar of the river still thundering in his ears. And then he broke the surface, chest heaving as he tried to get air in his lungs, and saw that the course of the river had swept him towards the far bank, where the leafless branches of trees hung down almost to the water's edge.

He lunged towards them, his right arm thrown out beyond

his head, hand grasping fresh air in a desperate last bid to catch hold of something. Anything. And he felt the rough bark of a low-hanging bough shred his palm. He closed numbed fingers around it, unaware that he had actually caught it until his shoulder was very nearly yanked from its socket. Unable to stop his forward momentum, the branch dipped and bowed as it fought against the flow of the river, and threw him side-ways to smash hard into the slope of the riverbank. He let go and clutched at clumps of grass and rock embedded in the embankment. He was out of the water and trying desperately not to slide back in. His legs were like lead weights as he tried to crawl further from the torrent snapping at his heels. Until finally he felt secure enough to roll on to his back and bark at the sky, lungs desperate to fill and refill and feed oxygen to his body. He pulled himself up on to one elbow and looked back across the river. There was no sign of the masked man. He was almost certainly gone, swept out into the loch.

Now Brodie started shivering. Uncontrollably, as his body tried to generate heat. But it was a losing battle, and he knew he would never make it back to the hotel. Almost centimetre by centimetre, he dragged himself up the bank, getting finally to his knees and crawling the last metre and a half up on to Lochaber Road.

Almost immediately he was blinded by the lights of a large vehicle coming off the bridge and heading towards him. He raised a feeble hand to shade his eyes and heard the hiss of brakes as the vehicle came to a stop. Then a man was crouching beside him, strong hands helping him to his feet. Above the howl of the wind, Brodie heard his voice: 'For Christ's sake, man, what the hell happened? You're soaked to the skin. You'll freeze to death out here.'

With an arm around his shoulder, he supported Brodie's failing legs to help him towards the passenger side of the truck. And Brodie saw then that this was a snow plough.

The cab was toasty warm and Brodie felt himself propelled into the passenger seat, barely conscious. Then the driver was beside him on the other side, a big man with a cloth cap and silvered whiskers that caught the light of the courtesy lamp. 'You need a doctor, man.'

But Brodie shook his head. Through chittering teeth that he could barely control, he told the driver that he only needed to get back to the International Hotel. It could be no more than a few hundred metres away.

The driver exhaled his exasperation. 'You're mad, fella. I'm going to the power plant at Ballachulish A. There'll be snow later, and we'll have to keep the road clear. But there's a duty doctor there.'

The words fell from Brodie's mouth like marbles from a jar. 'Just . . . just to the h . . . hotel.'

The driver took his snow plough right up to the front door, pulling in behind Brannan's SUV, and with chittered thanks, Brodie fell out into the snow. He was only vaguely aware of the plough reversing away as he staggered up the steps to the door and almost collapsed into the hall.

It was fully dark outside now, and the lights were on in the hotel. It was warm here, and Brodie stood for a minute, supporting himself with a hand on the wall, to try to catch his breath. 'Brannan!' His voice sounded inordinately feeble in the vast silence of the hotel. 'For fuck's sake, Brannan!' Still nothing. So much for waiting in for a call. With a great effort, Brodie pushed himself away from the wall and staggered to the stairs, using the banisters to pull himself up one step at a time.

When he reached his room, he was spent, hardly able to prise himself out of his wet clothes with hopelessly trembling fingers. He made it naked to the bathroom, flesh turning almost blue, and very nearly fell into the shower. It seemed almost impossible for him to turn the taps, but eventually he managed to start a stream of hot water tumbling from the showerhead, and he slid down to sit in the shower tray and let it cascade over his head and shoulders.

He could not have said just how long he sat there in that stream of hot water, but very gradually the feeling returned to his body, and with it, pain. Aching pain that seemed to infuse every muscle, every joint. And he reflected on how extraordinary it was that the icy waters of the tailrace and the river had so nearly taken his life, while the hot water that rained on him now from the shower was restoring it.

Finally he found the strength to get back to his feet, and stepped out to towel himself briskly dry. He wiped the steam from the mirror, and the face that stared back at him was bruised and battered from his encounter with the walls of the tailrace. Everything was stiffening up now, and he knew he needed to keep moving. He staggered painfully back to the bedroom and changed into his only remaining dry clothes. Clothes inadequate to protect him from the weather that powered unremittingly up the loch towards the village. He heard the first hail crackling against the window, and saw his reflection in it bulge with the force of the wind. With fingers that felt like bananas, he pulled on a pair of shoes, and searched through his sodden North Face to retrieve the Geiger counter zipped into an inside pocket. He had no idea if it would still function, but he wanted to take it to his meeting with Jackson to ask if he knew why Younger would have had it in his car.

He picked up the iCom earbuds that he had discarded on the floor and wondered if they had survived their underwater ordeal. He worked them back into his ears and asked iCom to connect him with the duty controller at Pacific Quay. Nothing. Either they had succumbed to the waters of the tailrace or the batteries were out of juice. He found the protective case that contained his glasses and took out the charging cable. After connecting the parts, he set it charging on the dresser. A winking green light offered the hope that it might actually still be working, and he headed off downstairs in search of something to eat, and more importantly, something hot to drink. He needed to warm himself up from the inside, too.

In the kitchen he found a coffee maker and brewed a tall mug of piping hot coffee, sweetened with several teaspoons of sugar to try to restore some of his energy. In a frying pan he cracked open several eggs he found in the fridge, fried them in butter, and sat down at the table to wolf them down. Between the coffee and the eggs, he was starting to feel vaguely human again. And his thoughts returned to McLeish. That he had killed both Younger and Sita seemed undeniable now. Though Brodie had no idea why. And the fact that McLeish had almost certainly been swept out to his death in the loch meant that the only person left who could throw any light on it all was Jackson.

He checked his watch. At least it was still working. It was almost time to leave for his rendezvous with Younger's contact. He stood up as the kitchen door swung open and a harassed-looking Brannan hurried in. 'Where have you been, Mr Brodie?' he began, before his voice tailed away and his eyes opened wide. 'What happened to you?'

And Brodie realised he must look even worse than the vision which had greeted him in the mirror. He said, 'Getting swimming lessons.' And as consternation creased Brannan's face, added, 'More to the point, where the fuck have you been?'

'Trying to find you.'

'Why?'

'She's gone.'

'Who's gone?'

'Dr Roy.' He jabbed a finger towards the door to the ante-room. It stood ajar, revealing darkness beyond. 'I locked that door after you'd gone this morning. Just to be safe, because I had to go into the village for some provisions. Then this afternoon, after you'd left again, I thought I'd just check.' He paused breathlessly. 'The door wasn't locked. Someone had forced it. And . . . she was gone.'

Brodie pushed past him and into the anteroom, reaching for the light switch. The lid of the cold cabinet had been lifted, and leaned back against the wall. The cabinet was empty.

'What do you think?' Brannan said.

Brodie turned back towards him. 'I think someone's fucking with us.' And he held out an open hand. 'I've got to go. Give me your car key. And I'm going to need to borrow a waterproof jacket.'

CHAPTER TWENTY-FOUR

He seemed to be driving headlong into the gale. Hailstones flew out of the darkness like sparks, deflecting off the windscreen. The outside temperature displayed on the dash of Brannan's SUV was minus two. He could barely see the road ahead of him, hail blowing around and drifting like snow on the recently cleared tarmac.

It took him longer than the ten to fifteen minutes predicted by Brannan. Several times he stopped to consult the map that lay open on the passenger seat, and to try to identify landmarks in his headlights. Finally he spotted the lay-by that Brannan had marked with a red cross on the map, and he pulled in off the road.

He sat for a while, steeling himself to face the storm outside, summoning his last reserves of energy, and felt the vehicle buffeted by the wind. The door was nearly whipped from his grasp as he opened it, and he had to battle hard against the wind to close it again.

He pulled up the hood of Brannan's anorak, and slipped the elastic of his headlight around it. Now at least he could see where he was going, hail slicing through its beam almost horizontally as he clambered down off the road to stumble through trees and a tangle of dead ferns towards the loch somewhere unseen ahead.

He very nearly ran straight into the bunker as it loomed suddenly out of the dark – a concrete pillbox that stood almost three metres high, just beyond the line of the trees and within sight of the water. It was hard to imagine a more inhospitable time and place to meet anyone. He felt his way around the walls to the front side facing the loch. A heavy steel door stood partially open, and electric light angled out from behind it towards the shore.

Brodie ducked inside, grateful to be out of the wind and the stinging hail whipped in on its leading edge. A single round LED light set into the roof cast a harsh yellow glow around the concrete walls, and the closed doors of what looked like an elevator.

He recognised Jackson immediately from Brannan's description. Tall, gangly, wiry ginger hair spraying out from beneath the hood of his parka. His face was the colour of ash, and nervous green eyes darted from Brodie, to the outside dark and then back again.

'Jackson?' Brodie asked unnecessarily.

The other man nodded. 'I don't want to be involved in this.'

Brodie said, 'Mr Jackson, you're involved whether you like it or not.' He tipped his head towards the elevator doors. 'Where does the lift go?'

'More than half a kilometre down a lead-lined shaft to the deepest level of the storage tunnels below. It's designed for escape rather than entry. Though those of us with security clearance have access badges on our key rings.'

'You didn't come up in it, then?'

'Good God, no. It wouldn't be safe down there.'

'Why not?'

Jackson rubbed his face with spindly white fingers. 'Look, I only ever spoke to Mr Younger on the basis of complete anonymity.'

'He was a journalist, Mr Jackson; I'm not. And if you don't want me to arrest you for his murder, I suggest you start talking. And fast.'

Indignation exploded from wet, purple lips. 'I didn't kill him! Why would I kill him? Jesus Christ, you can't be serious.'

'Then who did, and why?'

'I've no idea who.' He hesitated. 'Someone who didn't want him publishing his story.'

'And what story would that be?'

Jackson shook his head in slow desperation. 'I can't.'

And he wasn't prepared for the force with which Brodie banged him up against the wall. The policeman breathed in his face. 'My pathologist was murdered yesterday. And someone tried to kill me today, Mr Jackson. If you don't tell me what's going on here . . .' He didn't need to frame the threat in words. Its implication was clear enough.

Jackson shook himself free of Brodie's grasp. 'Okay!' He

almost shouted. He straightened his parka and breathed deeply, trying to figure out where to begin. Finally he said, 'Do you remember a story in the media about six months ago? It was on the radio and TV. An earthquake in the West Highlands.'

Brodie shrugged. 'Vaguely.' He thought about it. 'But the only reason that would have made the news is because you hardly ever get earthquakes in Scotland. And, as I recall, this one wouldn't even have made rings in a cup of tea. So no one made very much of it.'

'No, they didn't. But they should have.'

Brodie frowned. 'What do you mean?'

'It was a shifting of the tectonic plates on either side of the Great Glen Rift. Not far north of where we are now.'

Brodie stuck out a lower lip. 'Great Glen Rift? I've no idea what that is.'

'It runs roughly in a line from Fort William to Inverness, Mr Brodie. Effectively along the length of the Caledonian Canal. If you look at Scotland from space, it appears divided along that line into two parts.' He paused. 'Well, actually, it is. Sort of. And six months ago, the plates on either side of that divide shifted sideways. It wasn't a huge movement, and there wasn't that much felt above ground. But . . .' he shook his head in hopeless despair, 'there were fractures in the bedrock on both sides. Deep down.'

An unthinkable realisation began to dawn on Brodie. He pointed towards the floor. 'You mean down there?'

Jackson nodded. The ashen hue of his face was touched now by a green that almost matched his eyes. It spoke more than anything he could have put into words. He said, 'You know how the waste from Ballachulish A is disposed of?'

'Not in detail. Only that tunnels were excavated five, six, seven hundred metres down to store the stuff.'

Jackson screwed his eyes shut for a moment before opening them again to stare wildly at Brodie. 'We borrowed the idea from the Fins. You drill half a mile down into the bedrock, and excavate tunnels that fan out into a network of galleries. Radioactive waste from the reactor is put into boron steel canisters, which are then enclosed within corrosion-resistant copper capsules. Individual holes are drilled in the galleries. The capsules are placed into the holes, and then backfilled with bentonite clay.' He paused to draw breath. 'A permanent solution. The stuff is entombed forever. No further human or mechanical intervention is required, because the waste is now one hundred per cent inaccessible.'

Brodie thought about it. 'There must be a limit, though, to how much stuff you can put down there.'

'Of course. But there's enough capacity to store waste from the plant until 2120, when they'll seal it permanently and Ballachulish A will be decommissioned.'

Brodie said, 'And nobody foresaw the possibility of an earthquake?'

The shaking of Jackson's head was laden with sadness. 'That's just it. They did. In the early stages, the Scottish

Government commissioned a feasibility study into the whole waste-storage plan. The final study included a report which outlined the possibility of damage if there were any tectonic shifts in the Great Glen Rift. It did make it clear that such a thing was highly unlikely. The remotest of possibilities, Mr Brodie. I mean, almost certain never to happen. But, still, in the greater scheme of things, not impossible.'

'And they ignored it?'

Jackson's purple lips were tinged with white as he pressed them together in a grim line. 'Not exactly.'

'Well, what exactly?'

'If you look at the records in the government archive, Mr Brodie, you'll not find that report. It's not there.'

Brodie let disbelief escape from his lips in a breath. 'They buried it.'

'An inconvenient truth. Any further investigation into the possibility of tectonic shifts, or the damage that might result, would have taken years. The whole Ballachulish A project would have been put on ice. Might never have happened.'

The silence that fell between them then was broken only by the sounds of the storm raging outside and the wind that whistled in the half-open door and blew about their legs. The enormity of what Jackson had just told Brodie was slowly sinking in. But the reactor operator wasn't finished.

'The energy minister responsible for driving the whole nuclear project through the parliament in Edinburgh in the thirties gambled everything on Ballachulish A. It was going to be Scotland's energy future. And it was the rock upon which she built her whole career.'

Brodie looked at him. 'She?'

'The first minister. Sally Mack. Hoping now that the great Scottish voting public are going to re-elect her, forever grateful that power in Scotland doesn't have to be rationed like it is in so many other parts of the world.'

Brodie said, 'So she doesn't want this coming out before the election. In, what . . .' he checked the date on his watch, 'less than a week from now.'

'And with good reason. If it was revealed that she deliberately concealed a report warning of exactly what has happened, it would sink both her and her government.'

'And what exactly has happened?'

Jackson's breathing was shallow now as fear devoured oxygen and energy. 'No one knows for sure, Mr Brodie.' He steeled himself to say it. 'But radiation is leaking from the tunnels. A lot of it.' He raised his eyes towards the ceiling as if in silent prayer to make it all go away. Then refocused on Brodie. 'We figure that a fracture in the bedrock somewhere down there has damaged some of the boron steel canisters.'

Brodie frowned. 'Surely a radiation leak would trigger an alarm system of some kind?'

'Oh, it has. There's a team of experts here, combing the tunnels in radiation suits, trying to track down the source of it. A full investigation. But it's all hush-hush. Kept under wraps for reasons of "national security".' The sarcasm with which he imbued the words *national security* was not lost on Brodie.

And Brodie said, 'National security being another way of saying political convenience.'

Jackson sighed heavily. 'I don't want to get into the politics of it, Mr Brodie. But, well, call me a cynic. I figure that the whole fiasco will be cloaked in *national security* at least until after the election.'

Brodie said, 'How bad is it down there?'

'It's bad. A large section of the tunnel network has been sealed off to try to contain it.' He buried his face in his hands as if he could hide behind them. 'Oh, God,' he said, his voice muffled by them. And when he took them away again, Brodie saw tears in his green eyes. 'It's starting to leak out into the environment. This whole area shows readings way above safe levels.'

Brodie immediately thought of Addie and Cameron and felt sick. His investigation had turned into the worst kind of nightmare. Like a dream that haunts you during dark, troubled nights, then lingers long after the sun has risen. 'How was Younger going to prove all this?' he asked.

'I don't know how he managed it,' Jackson said, 'but he'd got hold of a copy of the disappeared report. Signed off by Mack herself. And he wanted to take readings. God help me, but he persuaded me to let him down into the tunnels. I told him radiation levels were probably fatal, but he was determined to go down anyway. He said he wouldn't be exposed for long. He needed the proof.'

And suddenly it dawned on Brodie what Younger was doing on Binnein Mòr. He remembered the radiation sensor just a little further along the ridge from Addie's weather station. Younger had wanted to take a reading from it. Crossing every 't' of his story. And if he had been fatally exposed to high levels of radiation himself, that would explain why he didn't want to waste half a day or more taking the long way up to the summit. Every minute counted.

'His piece in the paper and on the internet, Mr Brodie, was going to blow this government clean out of the water.'

The words had barely left Jackson's mouth when Brodie saw his head almost dissolve in an explosion of blood and bone and brain matter, throwing his body back against the far wall. The roar of gunfire in the confined space was enough to burst eardrums, and Brodie could hear only a loud, insistent ringing in his ears as he saw Jackson slide down the wall, leaving a bloody trail on the concrete. He turned as a shadow loomed in the doorway and caught only the briefest glimpse of a ski mask. Pain and light filled his head, then, consciousness sucked like matter into a black hole, and nothing remained but darkness.

The first thing he became aware of was a sensation of slowly sinking. Then came the return of pain filling his head, and when finally he opened his eyes, it was to be blinded by light. He could not immediately identify the source of the light. It seemed hidden by the square of ceiling above his head, and leaked out on all four sides. He was lying on a rubberised floor, half propped up against a stainless steel wall. He seemed surrounded, in fact, by reflective stainless steel. On the wall opposite, at roughly chest height, two illuminated buttons were mounted on a steel panel. The numeral 'one' above the numeral 'zero'. A ring of green light surrounded the 'zero' button.

Through the cloud of confusion that accompanied the pain in his head, it very slowly dawned on him that he was in the escape elevator, descending into the storage tunnels from the pillbox where he had met with Jackson. The sinking sensation was the slow downward motion of the lift. And all he could hear above the ringing in his ears was Jackson's voice saying *I told him radiation levels were probably fatal.*

Slowly, he got first to his knees, and then, with an effort, to his feet. He leaned a hand against one of the elevator walls to support himself. Lead-lined, Jackson had told him. The lift shaft was lead-lined. So for the moment he was protected from the radiation below. He staggered to the illuminated buttons on the far wall and jabbed his thumb at the 'one' button. The elevator continued on its slow but relentless descent. He jabbed it again, several times. Then, just for good measure, tried the 'zero' button. Neither had any effect. Panic started rising in his chest, and he pressed himself against the back wall, willing the elevator to stop. And still it persisted in its unrelenting downward passage. He closed his eyes. The doctor had given him six to nine months, which had seemed like nothing at all. And now they seemed like an eternity, and felt like life itself. Precious.

He could hear his own breathing in the confined space. Almost imagined he could hear the rapid beat of his heart, but really it was just the pulsing of it in his neck.

And when the lift came to a softly juddering halt, he held his breath, aware of the silence. It felt as though an eternity passed before a deep clunk preceded the opening of the doors.

He was not sure what he had expected. But death did not rush in to greet him. At least, not that he could see. Just warmth and light. Beyond the doors a cavernous cathedral rough-hewn out of the bedrock opened up before him, walls lined with pipes and trunking. It was well lit, bright lights reflecting off a polished concrete floor. The air was suffused with a soft electric hum, the source of which was not immediately apparent.

Brodie stood without moving for several minutes, imagining that his invisible enemy was killing him, even as he breathed it in, even as it was absorbed by his skin, and entered his body through every cut and graze. And yet he felt nothing. Smelled nothing but the acrid dust of drilled rock. And he wondered if the odd inflammation that Sita had found in Younger's lungs, and the sloughing of mucus in his intestine, was the result of radiation sickness. The samples she took would have revealed the truth back in the lab, but he figured they were gone now, along with the pathologist herself.

An unexpected calm descended on him. He was going to die anyway. And maybe those precious months would only have been an endless cycle of chemo and radiotherapy. A living nightmare. Better, perhaps, to die sooner. But not before he got out of here to settle the score. To get his daughter and grandson as far away from this place as possible. To bring the people responsible to book. To drop them to their fucking knees.

He pushed away from the back wall of the elevator and stepped out into the vast arc of this underground cathedral.

The main entrance into it was closed off by a large black door, perhaps five metres square, delineated by red light strips that cast a faint pink glow around the whole cavernous space. Brodie's footsteps echoed in the softly humming silence as he walked across the floor to examine it. There seemed no way of opening it from the inside, and he thought that the door itself was probably made of lead, immovable by anything other than some very heavy industrial mechanism. Huge tunnels fed away from the main space like spokes in a wheel and disappeared into darkness. A number of them were sealed. Rubber tyre tracks on the floor led off into others.

Brodie fumbled in his pocket for Younger's Geiger counter. He found a switch on the side of it and turned it on. The grey screen flickered to life, and immediately the device began to issue a piercingly high rate of audible clicks that fired through him like the pellets of a shotgun. Brodie had no idea exactly what level of radiation was being registered, but he had sat through enough movies to know that this sound was not good. The reading on the screen meant little to him either, and he quickly turned it off. The relief from the crackling was instant.

Better not to know, he thought, and pushed it back into his pocket.

He looked around now. There had to be another way out. These tunnels ran for kilometres underground. Surely there would be another escape elevator?

He crossed the hall and entered the nearest tunnel, reaching up to find that his headlight was still in place and still functioning. It pierced through the darkness that lay ahead. He saw lights running along the arc of the ceiling overhead, but had no idea how to turn them on. He set off, following the trunking that lined the wall to his right. Smaller tunnels fed away to his left at regular intervals. Again, some of them had been closed off. He passed a large electric trolley that appeared to have been abandoned and saw a red light somewhere up ahead. When he reached it, he realised that it was set high on the wall above another square door that, this time, stood open. Its delineating light strips were powered off.

Brodie stepped through it into a larger chamber, turning his head to direct light around its chiselled walls. It reflected back from a sign of red letters on a white background. EMERGENCY EXIT. And an arrow pointing off into darkness. He pressed ahead.

He had difficulty breathing now. The heat was suffocating, and he was perspiring down here when the world above was being plunged into an Arctic chill by an ice storm. It was very still in the tunnels. Almost peaceful. Why would he even want to escape back into the raging storm? He was so tired. All he wanted to do was sit down with his back to the wall and close his eyes. And maybe never wake up. And then he thought of Addie, and Cameron, and knew he had to keep going for them.

He walked on, past yet another sign, before the dark walls of a lead-lined shaft rose out of the floor to vanish into the roof space above. There was a single illuminated button set into the stainless steel to the right of the door, a ring of green light around it. His mouth was dry. He pressed it, and the door slid open, spilling bright yellow light into darkness.

He stepped into the light, and with a trembling finger pushed the 'one' button. If it did not respond, then these tunnels could well be his final resting place. His tomb. He might starve to death, or die from radiation poisoning before anyone found him.

To his relief, the door slid shut, and with the softest of judders, the elevator set off on its long, slow climb back to the surface.

The lift travelled at little more than walking pace, and took nearly ten minutes to reach the surface. Brodie stood leaning against the back wall with his eyes closed, trying not to think. After all, he wasn't out yet. He found himself transported into what felt like an almost Zen state of mind. Nothing mattered. Nothing existed beyond this space. All anger and sadness, all emotion, left him. Like spirits escaping after death. Minutes might have been hours, days or years. Time was irrelevant.

Then the elevator came to a sudden halt and the doors slid open. The cold was invisible, like the radiation, and it rushed in as the contamination he imagined he had brought up with him escaped. He opened his eyes, and the anger returned. A burning, all-consuming fury. He stepped out into the ice-cold of a concrete pillbox and put his shoulder to the bar that released the catch on the door. Heavy as it was, the strength of the wind outside caught it and flung it open. Brodie staggered out in the chaos of the storm and was nearly blown from his feet. Hail had turned to snow. Big fat flakes of it that filled the air and stung his face.

He could just make out the trees beyond the pillbox fibrillating wildly as they yielded to the wind. Perhaps twenty metres off to his left, the ground sloped steeply away towards the turbulent waters of the loch. If he kept the loch to his left and followed the shoreline, he would surely get back to the place he had met Jackson. He ducked his head and leaned forward into the wind, to thrust against it, forcing his legs to carry him through the snow, back the way he had come down below.

It was ten or more minutes before the concrete of the first pillbox reflected back at him from his headlight. He pressed himself against the near wall of it, taking momentary refuge from the power of the wind, then swung around to pull at the steel door. It was firmly shut and wouldn't budge. Whether or not Jackson, or what was left of him, was still in there was moot at this point. He was dead, and there was nothing Brodie could do to change that.

He wheeled away and staggered up through the trees, back

towards the road, hoping against hope that Brannan's SUV was still where he had left it. The glass of the passenger window caught and reflected the LED of his headlight as he scrambled up the embankment, and it was with huge relief that he felt his way around the vehicle, pulled the door open and almost collapsed into the driver's seat. It took an enormous effort to close the door again as the wind tried to rip it from its hinges. And then he was locked away in a bubble of comparative silence. The storm still raged beyond the glass, but it was muted now as it vented its anger, rocking the SUV on its wheels and obliterating its windscreen with snow.

Brodie sat for several minutes, gasping, fighting for breath, and when finally he took control again of his lungs, he avoided the rear-view mirror. He had no desire to look death in the face. He slipped the vehicle into drive, set the wipers to fast, and as soon as he could see out, swung the wheel hard around to head back to the village.

CHAPTER TWENTY-FIVE

Brodie left fresh tyre tracks in drifting snow as he drove up to the steps of the International Hotel and pulled up sharply. He jumped out and ran up to the front door.

'Brannan!' he yelled into the cold yellow light of the hallway. But as had so often been the case over these last two days, the owner of the International was nowhere to be found.

Brodie climbed the stairs as fast as his failing legs would carry him, and burst into his bedroom. The place was in disarray. His laptop gone, his half-dried clothes strewn about the floor. But whoever had searched the room and taken his computer had left his earbuds still charging on the dressing table. Careless. Someone in a panic. The green light that had been winking when Brodie left them was no longer in evidence. He lifted the buds and inserted one into each ear. The protective case for his glasses was lying on the floor. He breathed a sigh of relief to find the glasses still within, and he slipped them on, feeling the magnets lock into place.

He closed his eyes and prayed that it would all still work.

He said, 'iCom, record audio and video.' And heard a voice in his ears. *Now recording*.

He sat down then, and stared hard into the lenses, and began his account of the evening's events, replaying the nightmare memories in his mind as if he were watching them scroll across a screen. He tried to recall everything. McLeish's gloves in the garage. The attack at the hydro plant, and falling into the tailrace. Then his meeting with Jackson, and the reactor operator's story of leaking radiation and buried reports. It was clear, he concluded, that McLeish had survived the ordeal in the River Leven, and followed him out to Ballachulish A for his meeting with Jackson, killing the latter, and sending Brodie down in the elevator to meet his maker in the contaminated tunnels below.

When he finished, he knew it wasn't enough. So much detail he had missed. But there was no time to refine it. That would have to wait. Right now his only focus was on finding McLeish, and stopping him before he killed someone else. 'iCom, send.' His voice sounded flat in the cold light of his deconstructed bedroom. He stood up. Snow blowing against the window clung to it, held by the force of the wind, obscuring the view out to the loch. He closed his eyes and felt himself swaying as he stood. So tired. All he really wanted to do was lie down on the bed and drift away. His sense of balance deserted him, and he opened his eyes quickly as he staggered and nearly fell, heart pounding. He had almost fallen asleep standing up.

As he descended the staircase, Brodie called Brannan's

name several times. But the hotel was still deserted. The fire the owner had lit earlier in the bar was dead. Brodie pushed through the front door and on to the steps where he felt the wind pile into him as it powered up the loch, intensity accelerated by the narrow confines of the fjord.

He jumped into Brannan's SUV, grateful for its residual warmth, and spun it around to head back down to Lochaber Road, peering forward into thick white flakes driving through his headlights, wipers smearing wet snow across the windscreen. Despite being an all-wheel drive vehicle, the SUV slithered about the road as Brodie pushed it to the limits of its grip. He collided with the low parapet on the bridge over the Allt Coire na Bà, and accelerated out of the bend, gathering speed along the straight stretch of road towards the river. Lights blazed in the police station as he passed it, before skidding right at the Tailrace Inn and turning into Riverside Road.

Curtains were closed against the storm in most of the houses here, occasional cracks of light leaking out into the night. But cold, naked light flooded over the snow drifting on McLeish's drive where it had been so meticulously cleared away earlier in the day. The garage doors had not been closed. A light still burned inside. The curtains in the living room remained undrawn.

The pedestrian gate wouldn't open for the snow piled up on the other side. There were no fresh footprints or tyre tracks, and Brodie was the first to leave traces in the freshly fallen snow when he vaulted the gate and trudged through it towards the front door. There he hesitated. If McLeish was in the house, there was no telling how dangerous he might be. He had shot Jackson in the face, so he was armed.

He took a quick glance in the window, and saw Mrs McLeish sitting on the edge of her settee, leaning forward, hands clasped tightly on her knees. She was the colour of putty, rocking very slowly backwards and forwards, as if in a trance. Faraway and lost in another world.

He withdrew from the light, and circled the house through the snow, peering into every lit window. The kitchen was deserted. There was an unoccupied bedroom. The dining room simmered in soft light, but there was no one in it. No sign of McLeish. Finally he came around to the open garage and saw that McLeish's tools still lay strewn across the floor where he had left them. Nothing appeared to have moved since he was here earlier.

Brodie waded through the snow back towards the living room window and rapped softly on the glass.

Mrs McLeish was on her feet in an instant, looking hopefully towards the window. Then frowned when she saw Brodie's face caught in the light. She hurried out of the room and a moment later opened the front door. Her own face was a mask of fear, grotesquely shadowed by the light reflecting off the snow. 'What's happened?' she said.

'Is Calum here, Mrs McLeish?'

She frowned. 'He never came back from the hydro. And he's not answering his mobile.' She paused. 'Didn't you see him there?'

'Someone in a ski mask attacked me at the plant, Mrs McLeish. Tried to stove my head in with a monkey wrench.'

Her eyes opened wide.

'We both fell into the tailrace and got washed down into the river. I managed to get out. I can't say for certain what happened to the other fella.'

She shook her head in frightened disbelief. 'Well, it couldn't have been Calum. Couldn't have been! For God's sake, Mr Brodie, why would he attack you?'

'Maybe because he knew that I was on to him. That he had killed that missing journalist. Murdered my pathologist. But he must have got out of the river, because he shot a man in the face barely an hour ago.'

Her incomprehension was almost painful. 'What are you talking about? My Calum wouldn't hurt a mouse. Wouldn't do any of those things. I mean, why would he? Why?'

Brodie shook his head. It was the one thing that had been troubling him all night, niggling away in the furthest, darkest recesses of his mind. Motive. What possible reason could McLeish have had for any of it? And yet the evidence against him appeared irrefutable. The gloves, the paint on the bull bar. The doctored CCTV video. The attack at the hydro plant. But doubt was creeping in now. He said, 'I have no idea why. But he left the imprint of his gloved fist all over Charles Younger's face when he attacked and knocked him off the summit of Binnein Mòr.'

She was shaking her head vigorously. 'What gloves? What are you talking about?'

'Work gloves that he owns. They've got reinforced fingers with a distinctive pattern etched along the back of each one.'

It still made no sense to her.

'Here, I'll show you,' he said, and strode away towards the open garage door. Mrs McLeish folded her arms protectively around herself and followed him quickly, big wet flakes settling in her hair and on her sweatshirt as she ran after him through the snow.

Inside the garage, melting snow was pooling on the floor as Brodie strode across to the bench against the back wall. He lifted one of the tan and brown gloves and waved it at her.

'Seem familiar?'

Her eyebrows shot up in incredulity. '*Those* gloves?'

Brodie lifted the other one. 'Yes, those gloves, Mrs McLeish.'

The breath she expelled in consternation misted around her head. 'Those are his work gloves, Mr Brodie. He got them for working on the cars. But he found that they were good on the mountain, too. Great grip and protection for the hands. And kept them fine and warm.'

'Aye, well, your husband was wearing them when he followed Charles Younger up on to the mountain. And, I can assure you, they left their mark on him. A pattern that's too distinctive for there to be any doubt.'

Mrs McLeish crossed the floor and snatched one of the gloves from Brodie's hand. She looked at it, and then glared defensively at the policeman. 'Well, you should know that my Calum's not the only one to possess a pair of gloves like these.'

Brodie stared back at her, and felt all his certainty melting away like the snow on the garage floor. 'Who else?'

'One of the members of the mountain rescue team was very taken with them,' she said. 'So Calum bought him a pair for his last birthday.'

'Who?' Brodie demanded again. 'Archie McKay?' He remembered the team leader's pugnacity from the day before.

'No!' She looked at him as if he were mad. 'It was Robbie Sinclair.'

CHAPTER TWENTY-SIX

The fury had returned, and Brodie felt out of control. He saw the world through blood as he ploughed his way up the path to the door of the little police office. The wind whipped it out of his hand as he opened it, throwing it back to smash against the inside wall.

Robbie looked up, startled, from his computer, his battered face drawn and grey. His eyes opened wide as he saw Brodie framed in the doorway, snow swirling around him in the wind.

'You fucker!' Brodie shouted, and the younger man was out of his chair and vaulting over the counter almost before Brodie could draw another breath. The older man swung a fist, catching Robbie high on the cheek. But it wasn't enough to stop the other's momentum, and the two of them fell backwards into the snow outside.

Brodie felt punches falling about his face and shoulder, glancing blows struck in desperation. And he brought his knee up sharply to catch his assailant in the sweet spot between his legs. Robbie bellowed and rolled away, allowing Brodie to stagger to his feet, clutching at the door frame for support.

The snow blew into his face and into the office behind him, icy air displacing warmth from the house. But Robbie was back on his feet, too, growling like a wounded animal. And he charged again, sending Brodie crashing backwards into the police office. Brodie struck the counter hard and felt pain shoot up his spine. The strength of the younger man pushed him backwards across the countertop, and fingers like steel rods closed around his neck.

Brodie knew he was no match for the young policeman, but he also knew how to fight dirty. He caught Robbie's left ear and pulled as hard as he could, feeling soft flesh rip in his hand. Robbie screamed, immediately releasing his grip and staggering back against the door. It slammed shut behind him as he took a blood-covered hand away from his head, and Brodie saw that his ear was dangling by no more than a shred of skin. He propelled himself off the counter, and smashed Robbie up against the door, punching him in the throat, then smashing his forehead hard into Robbie's nose. He felt bone dissolve under the force of the blow, and warm blood splashed over both their faces.

A shrill voice cut across the sound of battle. Piercing, imperative. 'Stop it! Stop!!'

Brodie stepped back and turned to find Addie standing wild-eyed in the doorway to the house, a rifle raised to her shoulder, the barrel of it pointed at the two combatants.

'What the fuck?' she shouted, and took in the damage to Robbie's face and the ear dangling from it. 'Jesus Christ! What are you doing?' Wide eyes flashed from husband to father, and back again.

'Daddy!' Cameron's plaintive cry startled them all, as he squeezed between his mother's legs and the door frame and ran to his father, sobbing and shocked by the blood and shouting.

Before Brodie could move, Robbie had scooped his son up into his arms, wheeling away to snatch a letter opener from the counter. He backed up against the wall, the point of the blade pressed to the boy's throat.

There was a moment of incredulous silence in the tiny little police office, broken only by Cameron's frightened whimpering. He clung to his father, not for a moment believing that Robbie would do him any harm, but utterly discomposed by the conflict.

Addie gazed in disbelief at the man who had fathered her child. 'Robbie . . . what are you doing?'

'Don't do anything stupid,' he said. 'Put the gun down.'

She let the rifle fall a little from her shoulder, but held it in readiness to raise again should she need to. 'For God's sake, Robbie. Stop it. You're *not* going to hurt him. I can't *believe* you would hurt him.' She turned incomprehension towards her father. 'Why's he doing this?'

Brodie was still breathing hard. 'Because he's already killed three people, maybe four, Addie. And he doesn't see any way out for himself.'

Her lips parted, but there were no words.

Brodie said, 'He killed Younger. And he murdered Dr Roy when he realised that the skin she recovered from beneath the dead man's fingernails would reveal his DNA.' He turned to Robbie. 'Am I right?'

Robbie was breathing through his mouth as the blood began to clot in his nose. 'Couldn't let her check it against the database.'

'Because all serving police officers have to give samples of their DNA.'

Robbie swallowed.

'You'd already tried to set poor Calum McLeish up for it, just in case it all went pear-shaped. Doctoring the CCTV footage to make it look like he'd done it. And what did you do, borrow his pickup truck to push Younger's car into the ravine?' Brodie glanced at Addie.

Her face had set now. Disbelief giving way to anger. She nodded. 'Every time we had wood to pick up.'

Brodie looked at Robbie again. 'You knew there would probably be traces of paint left on the bull bars. Maybe even made sure there were. And then, of course, there were the gloves.' He let his eyes wander to Robbie's hands, then back to his face. 'Bet you destroyed yours. So the only pair would belong to McLeish. And if we went looking for a match . . .' He paused. 'So what did you do to the poor guy? Just one more body left in your wake?'

Robbie's face twisted itself into an ugly sneer beneath the blood. 'No need for a live fall guy any more, is there? Thanks to you.' He paused. 'McLeish'll burn now like everyone else.'

Brodie frowned. 'What do you mean?'

'You'll find out.'

Brodie stared at him for a long time, before he swivelled his head towards his daughter. 'He tried to kill me, too, at the hydro. And when he failed, he followed me to a meeting with Younger's contact from the nuclear plant. Pretty much blew his head off. Then bundled me into an escape elevator and sent me down into tunnels contaminated with deadly levels of radioactivity.' He turned back to Robbie. 'And yes, I probably did get a fatal dose of the stuff. I've no idea how much, or how long it takes. Maybe I've only got a day or two, who knows?' He forced an angry breath through his teeth. 'But what you didn't know was that I'm dying anyway. Fucking cancer. Dead man walking. All you've done is accelerate the process.'

Cameron's whimpering had subsided, and the boy clung to his father's neck, seemingly oblivious to the point of the letter opener pressed to his neck, or the tiny trickle of blood from where it had broken the skin. The child still seemed to have unwavering trust that his father would always protect him, no matter what.

Addie's rifle fell away as she went limp, at first with disbelief, and then despair. All three adults stood breathing the same air, sharing the same space, adding loudly to the same silence. Three lives in total disarray. Hope, belief, trust, all gone.

Addie's voice was very small when finally it was she who broke the silence. 'Why, Robbie?'

And he breathed his pain into the room, closing his eyes in distress. 'None of this was ever supposed to happen,' he said. 'They told me all I had to do was scare him.'

Addie said, 'Who's they?'

He scoffed. 'Oh, *they* don't have a name, or a face, do they? *They* send other people to do their dirty work. But they were going to ruin me. God, Addie, you know what a mess I'd got into. We were drowning in debt. They told me I would lose my job, my home, my family. All I had to do was this one little thing, and all my problems would go away. The slate would be wiped clean.'

Brodie said, 'Scare the shit out of Younger.'

He nodded.

'Only neither you, nor they, understood that here was a man who was prepared to risk radiation poisoning to get his story. He wasn't going to be deterred by a few verbals, or a handful of punches. You were always going to have to kill him.'

Robbie's head dropped to his chest. 'It all just . . .' He looked up. 'Got out of control. A total fucking nightmare.'

Brodie said, 'And there's nothing you can do now to fix it, Robbie. It's over. Give it up, for God's sake. Bring the nightmare to an end.'

Silent tears streamed through the blood on his face. 'An end for you, maybe. Not for me. Not ever for me.' He looked at Addie, with what almost seemed like an appeal for sympathy. 'I'm sorry, Addie. I'm so sorry.'

But there was no forgiveness in her eyes. He was still holding a blade to her son's throat. 'I never knew you at all, did I?' she said. 'All these years. A fucking stranger pretending to be someone I loved. Pretending to be someone who loved me.' Her eyes strayed to Cameron. 'Pretending to be someone who loved his son.'

'I do!'

'Then why are you holding a fucking knife to his throat?' Her voice reverberated around the room in anger verging on hysteria.

And as if he only now realised what he was doing, he suddenly threw the letter opener away across the office and unpeeled his son's arms from around his neck. Cameron's sobs of uncertainty returned as his father held the boy out towards Brodie. 'Here. He's your grandson,' he said. 'Take him.' And as Brodie clutched the boy to his chest, Robbie turned and fled out into the night, leaving the door swinging behind him in the wind.

Addie leaned the rifle against the wall and took two steps across the room to retrieve her son. Cameron flung his arms around her, and even in the midst of his distress and confusion, he turned his head towards Brodie and said, 'Are you really my grampa?'

Brodie felt his throat swell up as he fought back the tears, and was unable to find his voice in reply. He simply nodded, and the boy turned away again with the acceptance of a child who has understood nothing of what has passed between his parents and this man who was suddenly his grampa. But it was overwhelming, and he clung desperately to his mother and buried his face in her neck.

Addie stared in quiet desperation over her son's head at her father. Finally she said, 'What'll happen now?'

Brodie stepped across the room to close the door, then turned to face his daughter. 'Well, there's no point in me trying to go after him in the storm. He won't make it out of the village anyway. Not in this snow.'

She said, 'He has two hunting rifles. There was only one in the gun cabinet when I went to get this one.' She inclined her head towards the rifle leaning against the wall.

Brodie nodded grimly. 'Then he must have the other one stashed somewhere.'

'What do you think he'll do?'

'It's not so much a question of what he'll do, as what he'll try to stop me from doing.'

'Which is what?'

'Leave.'

'But you can't. Nobody can leave in this.' She lifted her eyes towards the world outside.

'No. But as soon as the storm is over, you and I and Cameron can fly out of here in the eVTOL. My guess is he'll wait till we try to make it to the football field, then pick us off.'

Addie was shaking her head, still struggling to come to terms with it all. 'He wouldn't. Surely to God?'

'He's lost everything, Addie. It's not a gamble any more. He has nothing left to lose.' He paused. 'If it's any consolation, I don't believe he'd hurt either of you. But he can't afford to let me get out of here alive. Not with everything I know.'

She closed her eyes, as if by shutting them she could somehow escape this waking nightmare. When she opened them again, she said, 'It would be crazy to try to get to the football pitch in the dark. Even if the storm was over.'

He nodded. 'Yes. Whatever happens, we should wait at least until first light.'

Almost as the word *light* left his mouth, the light in the tiny office was suddenly extinguished.

'Shit.' He heard her curse under her breath in the dark. 'Is that Robbie?'

'I don't know.' Brodie fumbled his way to the window and peered out into the darkness. 'It looks like all the street lights are out. Power lines must be down again.'

A match flared in the dark, and a flame flickered on the wick of a candle held in Addie's hand. Cameron sat wide-eyed on the counter as his mother shut the drawer beneath the counter and took him in her arms again. 'We're well prepared,' she said. 'This happens too often.' She handed him the candle. 'Come through and I'll light a fire. It's going to be a long wait.'

CHAPTER TWENTY-SEVEN

Brodie moved around the house like a ghost, drawing the curtains in each room before searching it by torchlight. Robbie was probably somewhere out there watching. He would be cold, and in pain, increasingly desperate. All of which would only make him more dangerous.

The double bed in the couple's bedroom was unmade, sheets and quilt tangled like detritus washed up on some barren shoreline.

This is where his little girl had spent all her sleeping hours with the man she thought she loved. Where they had made love. Where Cameron had been conceived. And he felt inestimably sad for her, knowing how it felt to be betrayed by the person you trusted most in the world. He wanted to go back downstairs and put his arms around her and tell her that everything would be alright. But he knew it wouldn't. Her life, and Cameron's, would never be the same again. Robbie had put an end to their future as surely as if he had put a gun to their heads and pulled the trigger.

The thought made him angry, but somewhere deep inside,

he felt just a grain of empathy for the man who had done this to them. For Robbie had also done it to himself. He was a lost soul, lingering somewhere in his own self-inflicted purgatory, before taking his final steps upon the road that once no doubt was paved with good intentions.

Brodie searched the wardrobe, rifling through drawers of underwear. He checked beneath the bed. Just the fluff and dust that had gathered unseen and undisturbed through all the years of their marriage.

He moved to Cameron's room, but found nothing there either.

In the bathroom, bottles of sedatives and painkillers crowded the shelves of a cabinet above the toilet. A jar of cotton wool balls. Flossers. Cotton buds.

Toothpaste and brushes sat in a tooth mug on a glass shelf above the sink. Used towels lay on the floor where they had been carelessly dropped by whoever last used the shower. Robbie, he thought, after dragging himself from the river, restoring life to frozen limbs by standing under hot water. Just as Brodie had done. In a laundry basket, he found all of Robbie's wet, discarded clothes, and wondered how he would have explained them to Addie.

But mostly he just felt despondent, bearing witness to the demise of a life, a relationship, a family. He knew just how painful that loss could be, and he ached for Addie.

Downstairs he went through every cupboard in the kitchen, even checking the fridge and the freezer, struck by the banality of everything he found. An ordinary life lived in expectation of an ordinary future. More children. Grandchildren.

He turned away and went back to the sitting room. It was filled with the soft orange glow of a wood-burning stove with glass doors. Addie's legs were tucked in beneath her as she leaned into the arm of the settee, Cameron's head in her lap. The boy was fast asleep.

While he searched the house, Brodie had heard her crying, every sob tearing at his heart, ripping through his conscience. But she was all cried out now, and sat puffy-eyed, gazing vacantly at the flames beyond the glass. He had no idea how much radiation people living in the village had been exposed to. There was no record of the reading Younger must have taken from the GDN radiation sensor at the summit of Binnein Mòr on the day he died. How much was safe? How much was dangerous? Brodie simply didn't know. But one thing was certain, he was not leaving here without his daughter and grandson.

He sat on the edge of the armchair opposite Addie, and vacant eyes flickered towards him. He said, 'Where did Robbie keep his stuff? Toolbox, climbing gear, things like that. In the garage?'

She shook her head. 'There's a wooden hut out back. It's pretty big. It was kind of like his den. I never went in there. Didn't want to. I'm sure he kept a secret laptop somewhere in it so that he could go online and lose more money without me knowing.' Her eyes filled with tears again. 'I should have, though. For so long I just ignored it, hoping it would go away. Burn itself out.' She scoffed at her own stupidity. 'Of course, I was just burying my head in the sand.' And the allusion brought back a moment of painful memory for Brodie, of his first meeting with Sita on the eVTOL. Her words filled his head. *Ostriches don't bury their heads in the sand. They don't hide from danger, they run from it.* Maybe that's what Addie had been doing, running rather than hiding. She said, as if she could read his thoughts, 'Maybe if I hadn't, this would all have turned out differently.'

Brodie said, 'You have no responsibility for any of this. You're a victim here. You and Cameron. Just as much as any of those people Robbie murdered.'

She dropped her head into her hands, fingers spread, and held the weight of it for a moment. 'I still find it hard to believe that he was really capable of any of this.'

In a strangely hoarse voice that made Addie look up, Brodie said, 'We're all capable of doing things others find hard to believe, Addie. Sometimes even ourselves.' And he wouldn't meet her eye.

He stood up.

'I'm going to go and search his shed. Where will I find the key?'

'It's hanging beside the kitchen door.' She hesitated. 'What are you looking for?'

'I'll know when I find it.'

*

He felt the force of the storm the moment he opened the kitchen door. The wind drove large flakes into his face as he stepped out in the darkness, and almost blew him off the top step. He waited for some moments in the hope that his eyes would adjust themselves to the available light. But there was no light, and he knew that he would never find the hut without the torch.

He had been reluctant to use it, knowing that Robbie could well be out there somewhere, just waiting and watching. And the light of a torch would offer a tempting target, caught in the sights of a hunting rifle. But surely even a man as desperate as Robbie would have had to take shelter from this?

Brodie tensed, taking a calculated risk, and the light from his torch raked across the snow-covered wilderness that was the back garden before it alighted on the hut away to his right. He made a run for it, crouching low, hindered in his speed by the depth of the drifting snow. At the door of the shed, he fumbled to get the key in the lock, taking far too long. Just waiting for the bullet in his back. And then he was inside, slamming the door shut behind him, and breathing a deep sigh of relief.

The beam of his torch fell across a cluttered workbench. Tools and cables, a soldering iron, a vice. There were boxes lined up on the shelves above it, all marked with their contents. Screws in different sizes. Nails. Washers. Nuts and bolts. On the floor beneath the bench, large plastic containers stood side by side, different coloured lids clipped in place.

He laid his torch on the floor and crouched down to open them. In the first he found a black laptop and dozens of printed gambling receipts, some dating as far back as four years. This, then, was Robbie's not-so-secret laptop. It was in here that he had sown the seeds of his own destruction.

In the second box Brodie found a silver laptop and a well-worn brown leather satchel. He opened up the laptop, but the battery was dead. He turned it over and saw a scuffed white sticker on the underside with Younger's name and address handwritten on it, left over from a repair in an IT workshop somewhere. He could only imagine what secrets the computer might give up when charged.

The satchel was stuffed with laser printouts – early drafts of Younger's story – and handwritten notes in an A4-sized ring-binder notebook. As in the notebooks he had found in Younger's glovebox, these too were in shorthand, with scribbled figures that meant nothing to Brodie. Then, from the rear division of the satchel, he pulled out a weighty document held together with a foldback clip. It was a poor photocopy of an original, but still clearly legible. Beneath the Scottish Government crest, the title on the cover page made Brodie catch his breath. RISK FACTORS IN THE AFTERMATH OF A TECTONIC SHIFT AT THE GREAT GLEN RIFT. It had a sub-heading. ASSESSMENT OF POTENTIAL SEISMIC DAMAGE AT BALLACHULISH A. So this was the report that Sally Mack had buried when she was energy minister in the thirties.

Brodie wondered why Robbie had kept all of this. Insurance,

maybe, against everything going wrong. Which, of course, it had.

He shone his torch around the walls and saw Robbie's climbing gear hanging from a row of hooks. Several parkas and thermal trousers, various telescopic hiking sticks, a couple of ice axes, and three different sizes of backpack. Beneath them, on the floor, a row of hiking and climbing boots and a box full of crampons.

Brodie selected the largest of the backpacks and lifted it down to start packing it with Younger's laptop and the satchel with its contents.

Against the far wall stood a row of cupboards with ill-fitting doors. He began pulling them open and dragging out their miscellaneous contents. Folders of old accounts, boxes of family detritus, the bits and pieces of a long-dead bicycle. And in the last of them, Sita's missing Storm trunk. He heaved it out on to the floor and opened it with trembling fingers. All the tools of the dead pathologist's trade, neatly packed away in fixed trays, and clips attached to the walls of the box. And in a sealed plastic bag, the jars and sachets of samples from Younger's autopsy, along with her notes.

Brodie stuffed them into the backpack and was about to leave when he spotted Sita's crime scene DNA analyser near the bottom of the trunk. He knelt down again to lift it out. He had no idea how it worked, and the battery seemed dead. He could get no read-out from its screen. But a paper printout, the result of her last analysis, curled out from a roll set into the back of the machine. He tore it off and straightened it out to run his eyes over the lines of print, and felt his heart push up into his throat.

When he returned to the kitchen, he slipped off Robbie's backpack and kicked the snow from his shoes. There seemed to be no let up in either the wind or the snowfall, and he stood for a moment with his back to the door breathing hard. Then he trailed the pack through to the sitting room where the fire was dying, but the air was still warm. Addie had drifted off, and Cameron was in a deep sleep, both breathing softly in the still of the room.

Brodie laid the pack against the side of the settee, lifted the rifle from the sideboard, and sat himself down in the armchair facing Addie. He laid the rifle across his knees, and turned the printout over in his fingers again and again, gazing silently off into space. He was startled by her voice.

'What did you find?'

He shifted focus to discover her watching him. 'Everything the *Scottish Herald* will need to put together the story Charles Younger was writing.'

'Which is what?'

And he told her what Joe Jackson had revealed to him in that cold concrete bunker on the edge of the loch. Her eyes opened wide in shock. Understanding for the first time, perhaps, the pressures that had been brought to bear on Robbie. Stakes that were too high even for him to contemplate.

They sat then for a very long time without saying anything, until at length he got up to chuck another couple of logs into the wood burner before resuming his seat. Sparks flew around inside it, funnelling fresh smoke up the chimney as the wood caught, and new flames sent light flickering around the room. He set the rifle once more across his knees.

She said very quietly, 'I can't imagine what it feels like knowing you are going to die.'

He glanced at her quickly, then away again. 'We're all going to die, Addie. Usually we don't know when, or how. Though in the early years, I think sometimes we believe we're going to live forever.' He drew a reflective breath. 'When the doctor first told me, it was like the biggest wake-up call ever. Fuck, Cammie, you're going to die! Who knew?' He sighed. 'It's a shock, and you feel sorry for yourself. Why me? Then, when that wears off, you start to get a perspective on it.'

He stared at the flames licking all around the logs.

'The hardest thing to come to terms with is the regret. I mean, life is an opportunity. The chance to do something that maybe won't mean much in the grand scheme of things, but will have significance in your own little universe. Which is not so little, by the way.' He looked at her. 'It's everything, Addie. It's your whole being. And my overwhelming sense of my whole being is one of failure. Of having somehow wasted my life. Thrown away that opportunity. Because, you know . . .' he shook his head, 'you always think there'll be time. To put things right, to catch up later. And there isn't. You waste your life on things that don't even matter. You want things you can't have, and dream of stuff that can never be. And all the time, your life is slipping away through your fingers, like so many finite grains of sand, squandered on . . . nothing. Then suddenly you're staring down the barrel of the end of your life, and all you're left with is the regret. The things you said, or didn't say. The things you did or didn't do. And it all seems like such a pointless fucking exercise.'

He forced himself to smile, a wry, self-deprecating smile.

'You know what's weird? I mean, the doctor gave me six to nine months. And I just had a dose of radiation that's going to cut that down to – who knows – days? Weeks? But I never felt so alive as I do right now.' He looked at her very directly. 'And never had a greater reason to live.'

The light from the fire reflected in her mother's eyes. She pursed her lips, and he felt her regret, too. 'Earlier today, when you told me it was Mum who'd had the affair, not you, you said there was more. But you weren't going to tell me then.' She paused. 'Now might be as good a time as any.'

And he felt the biggest regret of all filling him up, displacing everything else inside, like the cancer that was killing him.

CHAPTER TWENTY-EIGHT

2040

After the incident at the Leonardo Inn, I guess I was just sort of paranoid. Couldn't get those twin pictures out of my head – Jardine lying there on the floor of the hotel room, bleeding all over the carpet and swearing revenge, and Mel sitting in my car with tears streaming down her face, swearing she would never see him again. A lot of swearing going on, and I wasn't buying it.

I knew that without some kind of intervention by me, the two of them would be drawn together again, like magnets. Opposites attract, they say, and you couldn't get two more different people than Jardine and Mel. He was just a thoroughly bad piece of work – a ruthless, bullying, self-serving bastard. And Mel was one of the gentlest and most thoughtful people I'd ever known. And yet she couldn't stay away from him. Or he from her. I've never understood it; I never will.

So I set out to make sure that my worst fears would never be realised.

I was able to access his files on the police computer, read the reports of his parole officer. I wanted to know every last detail about the man. Where he lived, where he worked. Who his friends were, where he drank.

Would you believe it, he managed to get himself another flat in the tower block at Soutra Place. He must have liked it there. For my part, it was familiar territory. I knew where to park my car out of sight. Watch him coming, watch him going.

He started drinking in a pub on the south side where Celtic football fans gathered before and after games. The Brazen Head. He went to the home games every other Saturday. He was still driving the same banger he'd arrived in at the Leonardo, but he was careful never to drive when he was out drinking. Always taking a taxi, which must have cost an arm and a leg.

Not that money seemed to be a problem. He got his old job back at the bookie's, as if he'd never been away. He was getting on with his life, without a second thought for the mother and children he'd killed that night in Mosspark Boulevard. The only thing missing, it seemed – the only thing that would make his life complete – was Mel.

I don't know how long I followed him for. Must have been weeks. Every time I was off shift, every chance I got to head out east. Mel was very subdued all this time, and perhaps I should have been paying her more attention. I had no idea what was going on in her head, what kind of turmoil she was in. It was like when he was away in prison, he'd stopped existing for her. But now that he was out again, it was all that filled her waking thoughts. And maybe her dreams, too. At least, that's how it seemed to me. But like I said, I was pretty paranoid.

Anyway, it soon became clear to me that Jardine's weekly schedule included a rendezvous on the walkway running below the King George V Bridge. Tuesdays and Thursdays. It's an old bridge, the George V. Runs from Tradeston on the south side across the Clyde to Oswald Street in the city centre, just a spit away from the railway bridge that crosses the river at the same place. Nobody in their right mind would go down there at night, which is probably why it was an ideal spot for drug dealers to do business.

So Jardine had acquired another habit. Whether it was just dope, or something more upmarket like cocaine, I didn't know. I suppose I could have just tipped off the drug squad. Given them a time and place, and Jardine would have been caught in possession. A clear breach of his licence, and he'd have been back inside in a heartbeat. But he'd have been out again soon enough. Guys like him are no different from the cockroaches in that doctor's waiting room. Fucking hard to get rid of.

It was raining the night I followed him down to the walkway. Folk in the city were huddling under coats and brollies, so it was easy to stay anonymous. I kept a good distance, and watched him go down the steps and vanish into the dark beneath the bridge. A train rattled past overhead, and by its lights I saw the shadows of men moving around below the arch. I stayed out of sight, a good hundred metres away below the railway bridge, and watched as Jardine's dealers headed off in the other direction towards the Squiggly Bridge. Maybe they figured business wouldn't be good on a night like this. But, at any rate, they weren't hanging about.

Jardine came back along the walkway and up the steps towards me. He was walking slowly, despite the rain. He had no brolly and was getting soaked, more intent on protecting and checking the purchase hidden in his coat before heading back to his car. Satisfied at last, he pushed his hands in his pockets and began walking more briskly, almost as if he'd just noticed it was raining.

There wasn't another soul around. Well, there wouldn't be on a night like this. He was almost upon me when I stepped out to block his path just before the railway bridge. He got a fright, I could see that. Then, after a moment, recognition dawned and he relaxed. A smile spread across his ugly coupon, and I could see the gaps where I'd broken teeth, and the crooked turn of the nose that I'd busted.

'Well, well, well, if it isn't Mel's knight in shining fucking armour. What you doing here, Brodie? Gonna bust me? Didn't know you'd graduated to the narcs.'

I shook my head. 'No. Not going to bust you, Lee.'

'Oh, Lee, is it? Very familiar. What you looking for, then? A square go?'

I said nothing, and he must have seen the hate in my eyes as I fixed them on him in the rain.

'Oh, I get it. This is where you warn me never to cast a shadow anywhere near your precious wee bitch ever again.

That it, eh? Cos, let me tell you something, pal, there is fuck all you can do to stop me.'

Which is when something inside of me snapped. I mean, I'd come prepared. But I never really thought I'd go through with it.

He wasn't expecting it. A step towards him in the dark and the forward thrust of my right arm. I saw the surprise in his eyes as the blade slid up between his ribs and into his heart. But it didn't last for more than a second or two. He would never cast a fucking shadow anywhere ever again.

The weight of him fell forward into my arms, and I held him in a strangely grotesque embrace as I drew the knife away again. The lights of a train clattering south cast themselves on the dark waters of the river as I tipped him over the rail and watched him fall like a sack of tatties into the west-flowing currents of the Clyde.

He was gone in a moment. The same moment in which I realised just what it was I had done. How he had managed somehow to drag me down to his level, and lower. I threw the knife into the river as if the haft of it was burning my hand, and looked quickly around. But there was no one to witness my descent into hell. The same place to which I had just dispatched Lee Jardine. Not a living soul in sight. Cars rumbled by on the road bridge, headlights catching the falling rain. Folk on their way home, or out for the night. I looked down and saw Jardine's blood glistening wet all over the front of my coat. I stripped it off and rolled it up to tuck under my arm, and started running. Back towards the lights of the city. Back to the dark side street where I had left my car.

In preparation for the murder I never really believed I would commit, I'd stowed a roll of bin liners and a pack of hand wipes in the boot. I tore a bag from the roll and stuffed my coat into it, having checked the pockets for anything incriminating. Then I cleaned my hands on the disinfectant wipes I tore from the packet. There didn't seem to be any blood on my trousers or my shoes. But I wasn't going to take any chances.

It took me less than fifteen minutes to drive home to Pollokshields. I left the car in the drive and went in through the back door. Addie was out for the evening, and Mel had gone early to bed. I stripped off in the kitchen. Everything – shoes, socks, trousers, underpants – and stuffed them into another bin bag. Then I snuck upstairs to the guest shower room and stood under steaming hot water for a good five minutes, trying to wash away the guilt. Mel took up most of the wardrobe space in our room, so I kept my stuff in the guest room. I went in there to slip into clean clothes and tiptoe back down to the kitchen.

I threw the bin bag into the boot alongside the one with the coat and the discarded wipes, and drove west towards Paisley. It was somewhere on the Renfrew Road that I dropped the floor mat from the driver's side of the car, along with the bin bags, into a large wheely bin whose contents would be destined for landfill. And I sat in the car, eyes closed, drawing the breath that I had just robbed from another human being.

My heart was still hammering at my ribs, fit to burst, and all the regrets I would carry with me for the rest of my life came pressing in around me in the dark. The ghosts that would haunt me all my days.

I guess Jardine would have been missed when he failed to turn up for work the next day. Maybe he didn't show for a meet with his mates at the Brazen Head. But the alarm bells wouldn't have gone off full gong until he missed his first appointment with the parole officer.

I didn't know, didn't want to seem interested. It was only when Tiny told me one day that Jardine had gone AWOL and there was a warrant out for his arrest that I knew it was all going to come to a head soon enough.

I had no idea if he and Mel had been in touch in the time following the debacle at the Leonardo. I knew, or at least was pretty certain, that they hadn't met. But there must have been some line of communication between them, because in the weeks following that night under the George V Bridge, she became more and more withdrawn. If she had been expecting to hear from him, she must have wondered why he hadn't been in touch. Maybe she tried to contact him, I don't know. But the change in her was palpable.

I kept expecting to hear that they'd pulled him out of the Clyde. But it was almost three weeks before they did. Well downriver, near the Erskine Bridge. Of course, the body was decayed beyond recognition by then, but DNA identified him fast enough. The post-mortem located the fatal stab wound, and the traces of cocaine found in the pocket of his jacket led investigators to the conclusion that his murder was probably the result of a drug deal gone wrong. I knew there wouldn't be much effort made in trying to find his killer.

I suppose I thought then that I was home free. But it didn't really feel like it. I would never be brought to civil justice, perhaps, but natural justice has a way of finding you. There were other ways I would pay for killing that man.

I never told Mel about his body being taken from the river, or the results of the post-mortem. I was stupid to think that I could keep it from her. And sure enough, she heard. I don't know where, or how, but she did. Mentioned it to me one night at dinner, and I had to admit that I knew. I mean, she wouldn't have believed it if I'd claimed otherwise.

She seemed quite philosophical about it. Accepting, in that way of hers. As if she'd just heard that he was back in the Bar-L.

I really did think we were going to come through it, me and Mel. Until the night I got sent home early to find the cop cars and the ambulance in the street, and Mel dead in the bath.

CHAPTER TWENTY-NINE

2051

Addie's face, even from behind the wash of soft warm light that came from the wood burner, was paler than he had ever seen it. Her eyes were wide. He could see the shock in them.

Her voice came in barely a whisper. 'You murdered him.'

Brodie nodded, unable to maintain eye contact. He said, 'You always blamed me for the death of your mother. And you were right. Just not in the way that you thought.' He closed his eyes to focus on control of his breathing. 'Yes, I killed the man. And if I had my time over, God help me, I'd probably do it again. But what I know now, that I didn't know then, was that in taking his life, I effectively ended Mel's, too. In sliding that blade between his ribs, I might just as well have used it to cut your mother's wrists.'

If Addie had been shocked by his confession, it was possible that she was even more shocked now to see the tears that coursed down his face. Silent tears, a muted weeping that he choked back to swell in his throat and make his head pulse. Her eyes drifted down to the piece of paper that he had been turning over and over again in his fingers during the telling of his story.

'What is that?' she demanded, and he looked down to see how he had mangled the printout from Sita's DNA reader. He crumpled it up in his hand, and held it tight in his fist.

'Dr Roy had this new piece of kit,' he said. 'A portable DNA sampler capable of producing a reading at a crime scene.' He hesitated, his heart full of dread. 'I asked her to sample your DNA and mine.'

He felt her sudden fear reach out to him all the way across the room. 'Why?'

'I wanted to know if there was a familial match.'

'You said Mum told you—'

He cut her off. 'She did.'

'And you didn't believe her?'

'I did.' He paused. 'Ninety-nine per cent of me did. Probably because it's what I always wanted to believe.'

'And the one per cent?'

'Doubt. That tiny, shitty, niggling little grain of doubt that eats away at your soul until you just have to put it to bed. You just have to know.'

Her voice was very quiet. 'And now you do.'

He nodded.

'And?'

He pushed the crumpled piece of paper into his pocket

and forced himself to look at her very directly. It was almost painful to see the dread in her eyes. He said, 'Addie, I would never have told you about killing Jardine if I'd believed I was responsible for the death of both of your parents.'

325

CHAPTER THIRTY

Brodie had no sense of there being a moment when he drifted off to sleep, and he was startled to be awakened by the sound of a door opening.

Addie stood in the hall, kitted out as if she were intent on climbing a mountain. Her hair was tucked up under a dark blue woollen hat, and she clutched Cameron in her arms. He wore a parka and welly boots and mitts, and a cagoule that folded around his neck to keep him warm, his sleepy little face peering out from behind soft grey fabric.

Brodie blinked and realised there was light seeping in from beyond the curtains.

'The storm's passed,' Addie said. 'It's light enough for us to go.'

He struggled to his feet. 'You shouldn't have let me sleep.'

She said, 'Like you didn't need it.'

He struggled into the anorak he had borrowed from Brannan, and heaved Robbie's old weekend pack on to his back. He felt the weight of the laptop in there, and Sita's samples, and all the additional burden of responsibility that rested on his shoulders for getting his family out of here.

Addie held out a box of cartridges. 'Some additional rounds in case you need them.'

He took the box and stuffed it in his pocket. He said, 'Don't you think about anything except keeping Cameron safe and getting him to the eVTOL. Let me worry about Robbie.'

Outside, the snow lay thickly over everything, deep and unbroken, robbing the world of definition. There was light in a clearing sky, and in the absolute stillness that settled across the village and the mountains in the wake of the storm, all that could be heard was the dawn chorus. Birds emerging from wherever it was they had taken shelter, to greet the new day. Oblivious to the fear that stalked the streets of this tiny settlement.

Brodie stood in the doorway, clutching the rifle across his chest, scanning the rooftops and the near horizon. But the land rose steeply into the trees on every side, and Robbie could have been anywhere. He must have known that with the storm over, Brodie would take the opportunity of first light to try to make it to the playing field. He had every advantage.

Addie stood at her father's back, holding Cameron. She said, 'We're sticking close to you all the way. I don't think he'll risk a shot if there's a chance of hitting one of us.'

'Don't be stupid!' Brodie turned and growled at her. 'I'm not using my own family as human shields.'

She met his eye, unflinching. 'Dad, that's not your decision to make. We're in this together, or you go on your own.'

And again he saw something of himself in her. That wilful

stubbornness that so had characterised most of his adult life. He knew there would be no arguing with her.

The snow lay at least a metre deep, and more where it had drifted. So they made slow progress up Lochaber Road from the police station, huddled close, Brodie with his rifle held ready to raise to his shoulder. He scanned north and west, Addie raking keen eyes to the east. A plume of snow raised itself from the road just a couple of metres ahead of them, followed by the delayed crack of a rifle shot smothered by the acoustic muffling of the snow. A group of startled crows lifted black into the sky above a gathering of trees almost directly to the north, and Brodie swung his rifle in that direction to release a shot. He felt the kick of it against his shoulder, knowing that there was zero chance of hitting anything other than a tree.

Addie said, 'He's not that bad a shot. He's just letting us know he's there.'

Brodie nodded. The real test would come when they got to the football pitch.

It was heavy going through the snow with legs that ached and had to be lifted to make every step forward. Past Lochaber Crescent on their right. Mamore Road on their left. Not a single villager venturing forth in the aftermath of the storm. But curtains twitched with the sound of the shots, and unseen eyes watched them from behind glass that reflected only the glare of the snow. Semi-detached houses with upper dormers, and satellite dishes crusted in white. Fences barely rising above the drift.

The banks of the Allt Coire na Bà were smothered in shelves of compacted white crystals, the rush of black sparkling water beneath the bridge cutting a tortuous path beneath a skin of ice. The barrier that Brodie had clipped with Brannan's SUV the night before was lost beneath the snow.

Now the trees crowded in, close to the road, and rose steeply into the calm of the morning. It seemed inordinately dark among them, perfect for Robbie to move unseen, following their progress around the curve of Lochaber Road as it headed out of the village.

They passed a cottage among the trees on the other side of the fence to their left, snow clinging to the steep pitch of its slate roof, insulation sealing in its warmth, so that the snow would remain there all day, unmelted. Then, up ahead, they saw the turn-off into the tarmac parking area in front of the sports pavilion.

They smelled the blaze before they saw the glow of it in the sky and the smoke rising above the trees. But it was not the sweet smell of woodsmoke. It was the acrid stink of man-made materials, toxic, and belching abnormally black smoke. For a moment, Brodie thought that Robbie had somehow managed to set the eVTOL alight. But as they reached the turn into the football field, they saw Eve half buried under snow in the middle of the field where she had landed. The ground around her appeared undisturbed.

It wasn't until they left the road behind that they saw for the first time, beyond the trees, the flaming bulk of the

International Hotel. Across the valley, sirens were sounding at the fire station, but it would not be a simple matter to get here through metres of drifting snow. The flames licked high above the treetops, and Brodie understood for the first time what it was Robbie had meant when he'd said of Calum McLeish, *he'll burn now like everyone else.* Everything would be destroyed in the fire. Younger's corpse. And Sita. And McLeish. No doubt Robbie had intended for the evidence stored in his hut to go up in flames as well. He must have figured that Brodie had found it by now. Another reason he couldn't afford to let him leave.

The glow of the fire flickered orange across the white that lay thick on the playing field, ash falling from the sky like the snow before it. The prevailing breeze carried the smoke through the trees in their direction. But, even so, Brodie knew that as they crossed the field to the eVTOL they would be easy targets.

He glanced behind them, but there was no sign of life among the trees on the hillside. Addie stuck close to him, only too aware of the danger, as they ploughed across the snow-covered tarmac to the shelter of the pavilion. Brodie saw immediately that there was no green light on the reader attached to the plug. Either Eve was fully charged, or the cable had been severed again. He lifted the end of the cable, and began pulling it up from the snow as they crossed slowly towards the eVTOL. Halfway there, the severed end of the cable prised itself free of the snow where McLeish's repair had been ripped apart.

'Shit!' Brodie cursed and knelt down, searching in the snow

for the other end. They couldn't afford to leave a length of cable dangling from the e-chopper as it took off. *If* it would take off at all. He found it and ripped it up through the crusting white snow as they started to run towards Eve.

When they reached her, he waved his RFID card at the door and it slid open. He half turned towards Addie. 'Get the boy in, quick.'

Addie bundled Cameron unceremoniously into the back of the eVTOL, the boy protesting all the while at his rough treatment. Brodie pressed the return button inside the cable hatch, and the remainder of the severed cable began to reel itself in.

'Get in,' he shouted at Addie, almost at the same moment as the bullet hit him somewhere high on the left side of his chest. He spun away in a spray of blood to slam against the open door as the crack of the rifle shot rang out across the field. Addie saw the panic in his eyes.

This time it was she who shouted, 'Get in!' And she half lifted, half pushed him into the cabin of the eVTOL, before stooping to pick up the rifle that had fallen from his grasp. She turned around, back to the e-chopper, as Robbie started walking across the field towards them. The blaze of the International lit him orange down one side, and cast a long shadow across the snow. He had a scarf wrapped around his head, and held his rifle at chest height as he advanced slowly towards the aircraft. The air was filled with smoke and the crackle of flames. A siren wailing somewhere in the distance. Addie raised the rifle to her shoulder. 'Don't come any closer.'

He stopped then. A sad smile on his face. He shook his head. 'Addie, you know you're not going to shoot me. Just like you know I'm not going to harm you or Cameron.'

'Try me.' Her voice sounded bolder to her than she felt. She pressed the rifle harder into her shoulder, her finger crooked around the trigger.

He said, 'Addie, you don't have it in you. And, really, I mean you no harm.' He paused. 'But I can't let your Dad leave. I can't.'

'Take one step closer and I'll drop you where you stand.'

His smile became strained. 'There's too much of your father in you. Maybe I know now why you hated him so much.'

Without taking her eyes off Robbie, Addie half turned her head towards the open door. 'Dad, don't let Cameron see this.'

Inside, Brodie was almost numb with the pain. There was a lot of blood, and he could only just see Cameron's frightened eyes in the gloom. 'Come here, son,' he said, his voice the hoarsest of whispers. But it was invitation enough to propel the boy into his arms, and he turned his grandson away from the open door, wrapping himself protectively around the child.

Robbie's smile was gone now. 'You're making a mistake, Addie.'

'No, you're the one making the mistake, Robbie, if you think I'm going to let you kill my father.'

He stood for a moment, all humanity leached from his eyes - a man who had killed too many times. He lifted his rifle in a single, swift movement, and the shot that rang out spun him away, his fall cushioned by the depth of the snow. He half sunk into it, blood spreading quickly around him, rabbit fear in his eyes. As he tried to speak, blood gurgled into his mouth.

Addie stared in horror at what she had done. This was the man who had changed her life, persuaded her that she should make her future here with him, in this hidden valley. The man who had fathered her child. She had aimed for the largest part of the target, his chest, afraid that if she simply tried to disable him with a shot to the leg, she would miss, and he would shoot her instead. But, still, she felt sick to the core.

She became aware of Brodie yanking at her hood. 'Get in,' he whispered, still shielding the boy from the sight of his father lying bleeding in the snow. And as Addie climbed into the eVTOL with leaden legs, he closed the door, summoning all his energy to bark at Eve. 'Eve, initiate our return journey.'

Her voice came back to them. *Low battery, Detective Inspector. Range limited.*

'Just go,' he barked. 'As far as you can take us.'

After a moment she responded. *Flight initiated.* And the rotors above them began to spin, snow flying off in all directions, the power of the downdraft blowing it clear of the glass. The screen at the front of the cabin displayed a battery symbol in orange beneath the warning *RANGE THIRTY MINUTES*. It didn't matter to Brodie. He wanted to get them away from here. And thirty minutes flying time would just have to do.

With the gentlest lurch, Eve lifted herself up out of the snow and wheeled away, rising over the trees and the flames of the

International Hotel. Brodie peered through the glass, back the way they had come, and saw Robbie lying spreadeagled in the snow. Even from here he could see the blood.

And he saw the figure of a man running across the field to grasp Robbie's parka by the hood and start dragging his prone form through the snow towards the blazing building. Even as Brodie watched, the man turned his face up towards them. Brannan! It was fucking Brannan! And then realisation struck Brodie with sickening clarity. Brannan was the face of the faceless *they*. He was *their* man here. He had been pulling all the strings the whole time. Orchestrating everything. And now he was pulling Robbie into the fire, so that he too would go up in smoke.

But what made no sense is why Brannan would have let them get away. He could have disabled the eVTOL, set fire to it, just as he had done to his own hotel. Brodie shut his eyes and shook his head as a surge of pain took away his breath.

Addie took Cameron from his grandfather's grasp and held him to her, trembling almost uncontrollably. She glanced at her father as he eased himself forward and into the front seats, leaving a trail of blood across the leather. An involuntary cry of pain escaped him as he slid the weekend pack from his back, letting it fall to the floor. He dropped into the seat, his breathing laboured.

Below, the dark waters of the loch swept past, the unbroken white of the mountains rising up around them into a clearing blue sky. And as Eve lifted still higher, the early sun breached the peaks behind them, to send sunlight cascading west along the fjord, and filling the cabin with a golden light.

Still clutching Cameron, Addie manoeuvred herself into the seat beside her father and reached over to push her hat and gloves on to the wound beneath his anorak, pulling the drawstrings tight to create pressure on the wad where the bullet had entered. And another voice filled the cabin.

'Detective Inspector Brodie, this is air traffic control at Helensburgh. We're going to schedule a landing for you at Mull. You should have just about enough juice to get there. We'll monitor battery range remotely.' A pause. 'We have a video message for you.'

They were passing Glencoe village on their left now, and the power station at Ballachulish A two hundred feet below. Moments later they overflew the barrier bridge at the narrowest point of the loch and skimmed out across the open expanse of Loch Linnhe.

The monitor flickered and the battery symbol vanished, to be replaced by today's date, white letters on black, that in turn gave way to the battered face of Brodie delivering the report he had sent the previous night. Text scrolled across the bottom of the screen: *Report by DI Cameron on the death of Charles Younger*.

Brodie listened to his voice speaking. But they were not his words. His lips moved as if they were, but he knew he had never spoken them.

Dr Roy's post-mortem on Charles Younger, he said, had returned a verdict of accidental death. An apparent fall while climbing on Binnein Mòr. But the fire at the International Hotel, in which the pathologist had perished today, meant that her report and all her samples were lost.

He shouted at the screen, blood in his spittle. 'That's a lie! That's not me. I never said that. Younger was murdered. Murdered, for fuck's sake!' He lashed out and struck the windscreen of the cabin, his fist smearing blood on the glass.

Addie's voice, right beside him, was tiny and frightened. 'What's happening, Dad? What does it mean?'

He turned blazing eyes on her. 'It means I've been fucking had.' He fumbled in his pockets for his iCom glasses and snapped them in place with blood-sticky fingers. 'iCom, scan the video,' he shouted, focusing his gaze on the screen.

After a moment, his iCom returned its verdict. *Video genuine.* And a green *GENUINE* symbol flashed in his lenses.

He yanked the glasses from his face and threw them away across the cabin, and, with fumbling fingers, pulled the earbuds from his ears. 'Fuckers!' His voice reverberated around the cabin, his grandson shrinking into his mother, fear in the wide-eyed stare he directed at his grandfather. 'The software in these things isn't the latest version. *They* have the latest version.' He tried to bring his breathing under control. 'Eve, place a call to DCI Maclaren at Pacific Quay.'

He listened with dismay to the silence that greeted his request.

Addie said, 'I don't understand. What's going on?'

'That video of me . . .' Brodie waved his hand at the screen.

'It's not me. It's a deepfake. What do they call it . . . ?' He searched for the term. 'Neural masking.' He slammed his fist down on the dashboard. 'They've set me up for this.'

Eve interrupted. *RANGE FIVE MINUTES*. And an alarm began to sound. A piercing, repetitive wail that filled the aircraft. His video was replaced on-screen by a flashed warning: *BATTERY LEVEL CRITICAL*. The battery symbol was red.

They were over Mull now, and Brodie looked down in impotent frustration as they flew over the golf course above Tobermory, the land passing beneath them before giving way to the Atlantic Ocean sweeping in from the west in white-crested waves.

'Why aren't we landing?' Addie asked, fear making her shrill.

And he realised now why it was that Brannan had let them go. 'Because they're going to drop us in the ocean,' he said. 'You, me, Cameron and all the evidence of government cover-up at the nuclear plant. We're dangerous. And expendable.'

He stared out, wild-eyed, as they overflew the upper half of the island of Coll, and the vast expanse of the Atlantic ahead beckoned them to their final resting place. He had forgotten now about his wound. The survival instinct had kicked in, adrenaline overriding pain.

'No! No! No!' he bellowed. He stood up and swung a fist, feeling bones breaking in his hand as he struck the glass and smeared yet more blood on it. He turned to press his back to the windscreen, arms stretched out to either side like Christ on the cross. His mind was racing. Thoughts tumbling one over the other in blind panic. But he knew there was something he could do. Something just out of reach. If only he could remember.

He looked down at daughter and grandson staring up at him in sheer terror, and he could hardly breathe. He said, 'Jesus Christ, Addie. I was in the delivery room when you were born. I watched you draw your first breath. I'm not going to watch you draw your last!'

He slumped into his seat, burying his face in his hands. Think! Think! Think! It was so hard above the wailing of the alarm and Eve's constant prompting to buckle up again. And he knew at any moment that Eve was simply going to stop flying and drop them silently into the ocean. He tried to focus on the day he flew downriver to pick up his flight out to Mull. The technician in yellow oilskins who had run across the grass from the clubhouse at Helensburgh golf course. He had sat in beside Brodie and primed Eve for flight. Brodie screwed up his eyes. Tiny was so much better at this than him. He noticed things, remembered details that Brodie missed. A visual thing, he'd said it was. You could remember images better than words. And better still if you could link either to something personal. Something you could relate to.

Brodie tried to replay in his mind what it was that the technician had done. Of course! He'd tapped the screen simultaneously with his index and middle fingers. The image of him doing it returned from some deep memory recess that Brodie almost never visited. Twice. He'd tapped it twice. Brodie leaned forward to do the same, and the warning message was wiped away, to be replaced by the eVTOL's original welcome page. A horribly incongruous photograph of the aircraft taken on a sunlit day, and set against the clearest of blue skies. Brodie could remember thinking how unlikely it was that this photo had been taken in Scotland.

Absurdly, Eve addressed them as if for the first time. *Welcome to your Grogan Industries Mark Five eVTOL air taxi.* How had the technician responded, sitting there dripping rain from his glistening oilskins? Something had chimed with Brodie at the time. Something almost subliminal. The technician had identified himself with what was certainly his own unique code. Three letters and a three-digit number. Brodie could very nearly hear his voice. And then it dawned on him why the numbers had registered. His birthday. It was his birthday! Year and month, 496. April, 1996.

'Dad, we're losing height!' Addie's voice beside him was brittle with panic.

But he figured that had to be illusory. When Eve ran out of battery, her rotors would simply stop, and they would drop from the sky. He resisted the temptation to look and forced himself to keep thinking, trying to recall the technician's voice. But the warning siren was still filling his ears and it was difficult to think above it.

The man had used the NATO alphabet. Key words representing each letter for clarity. What were they? 'Come on, come on,' he urged himself, almost unaware that he was speaking out loud. And then he remembered how Eve had responded, calling him Zak. 'Z-A-K,' he said suddenly. 'ZAK496.'

He caught his breath to try to steady his voice and speak clearly.

'Zebra-Alpha-Kilo-496.' And then Zak had issued an instruction. What exactly had he asked Eve to do? *Activate remote.* That was what he had said. Brodie was sure of it. Now he had to ask her to do the opposite. He said, 'Eve, deactivate remote.'

And he was astonished to hear her respond immediately to the command. *Remote deactivated.* He breathed his relief, and heard the exhalation rattle in his lungs. Now Brodie had control. Not some bastard in a darkened room sending them to their deaths.

He said, 'Eve, turn around and put us down at the nearest safe landing point.'

The sunny photograph of the eVTOL vanished from the screen, to be replaced by a map of the Inner Hebrides. Their route across Mull and Coll and out to sea was traced in red, concluding in flashing circles of orange. A return route in yellow retraced their flight to the nearest landfall. The island of Coll, the chosen landing spot pulsing in circles of green.

They felt Eve bank left and turn through one hundred and eighty degrees, and the distant outline of Coll swung back into view. At the same time the eVTOL began losing height. No question about it this time.

Brodie hardly recognised his own voice. 'Eve, do we have sufficient battery?'

Battery life unknown. She sounded so calm. As if her programmers themselves had made no distinction between life and death.

Brodie felt Addie clutch his arm as Coll grew nearer. They were barely three metres above the waves now, fearing that at any moment they would fall into the brine. Salt spray blew back across the windscreen, blurring their vision. And still the alarm sounded.

They could see a beach, silver sands cleared of snow by an incoming tide. Beyond it, tufted machair land, sparsely covered by snow, was dotted with dozens of hardy, grazing, black-faced sheep. Beyond them, a road, a collection of huddled buildings, a farm.

And the rotors stopped turning, as silently as they had begun. The eVTOL dropped the final metre into snow and peat bog, landing heavily and turning on to its side, propelled forward by its own momentum.

It was chaos in the cabin, all three flung from their seats and sent sprawling as Eve slid across the snow on her side for another twenty metres, before coming to an abrupt halt against a line of broken fencing.

Cameron was wailing, in fear more than pain. Addie clambered over the upturned seats to grasp him to her, holding him close for just a moment before checking him for damage. But children are far less brittle than adults, and beyond a lump the size of an egg coming up on his temple, he seemed unhurt.

She turned around to see Brodie slumped at an awkward angle across the far door. He was looking at her across the space between as if it were some eternally unbridgeable gap. His breathing was laboured, and in his eyes she saw a look she had seen once before, when Robbie had taken her hunting and shot a deer. It had still been alive when they reached it, eyes full of incomprehension, but also accepting of death. And she had watched the light of its life die as Robbie pulled the trigger for a second time. She had never gone hunting with him again.

Now she scrambled across the upturned cabin, but he held out a hand to stop her. Beyond him, through the glass, she could see people running towards them from the farmhouse. He said, 'You're going to have to do this on your own now.'

'Don't be silly.'

She tried to sit him up, but he pushed her away. There was blood everywhere. 'Addie!' His voice was insistent. 'They're going to try everything in their power to stop you.' He fought to get more air in his lungs. 'So you're going to need help.'

CHAPTER THIRTY-ONE

Tiny sat slouched in his armchair, balancing a half-drunk beer on the arm of it. The television was on. News coverage of a political rally held by the Eco Party. The venue they had chosen was far too big for the number of its supporters crowding around the stage waving banners and saltires. The close-up shots made it appear full. But the TV director was showing his political bias by intercutting with wide shots revealing the emptiness of the hall beyond. It was an anticlimax to a bitterly fought campaign in which the Ecologists had gained almost no ground on the ruling Democrats, who were scheduled to hold a triumphant rally the following night on the eve of an election they were certain to win.

Tiny was paying it no attention. It was a distraction in the corner of the room, like the flickering flames of the living-room fires he remembered from his childhood. Sheila was sitting on the settee opposite, playing some word game on her tablet. They didn't talk much these days, drifting apart as they grew older, and without the glue of children to keep them together. But they were still comfortable with each other.

Tonight she had commented on how distant he seemed, coming home at the end of his shift to eat a carry-out pizza from a box on his knees. He had told her there was a lot on his mind. Just work stuff. She had never cared for Brodie, so he didn't really feel like telling her that his best pal had been killed in an air crash. It had been rumoured for a couple of days that his eVTOL air taxi had ditched in the sea somewhere off Mull. He had been shocked to the core to hear it. But no one had been able to provide confirmation. Not even the DCI. Until today. But the air taxi had not, it turned out, ditched in the Atlantic as first reported. It had crashed on the Isle of Coll, and they had pulled Brodie's body from it, killed by a bullet from a rifle. No one could quite believe it.

Tiny had spent most of the evening thinking about Cammie, remembering all their scrapes and adventures, and hoping against hope that somehow reports of his death had been greatly exaggerated. He knew, of course, it was a forlorn hope. When he'd first heard rumours about the eVTOL going missing, he had tried calling Brodie on his iCom, but the call had gone straight to messages. Which had not augured well.

Now on his third beer, he was subsiding into distant mawkish memories, and getting quietly emotional.

When the doorbell rang, it did not immediately penetrate his thoughts. It was Sheila's voice that woke him from his reverie. 'Who could that be at this time of night?'

Tiny looked up. 'What?'

'The doorbell.'

And right on cue, it rang again. Sheila clearly had no intention of answering it, so Tiny heaved himself out of his armchair to lay his beer on the coffee table before heading out to the hall to see who was there.

He turned on the outside lamp before opening the door. The rain that had been falling all day cut through the light it cast upon the steps and the path beyond.

A young woman stood on the top step, long auburn hair escaping from the hood of her parka, wet and smeared across her face. She was holding a child in her arms. A young boy who was fast asleep, his head resting on her shoulder. Tiny frowned. There was something oddly familiar about them both, but he was sure he didn't know them. 'Yes?'

She said simply, 'My dad told me you would help.'

CHAPTER THIRTY-TWO

Addie walked up Renfield Street in the rain. There was not a breath of wind beneath a bruised and ominous sky, and the large teeming drops raised a mist on the pavements and filled the gutters as they ran in spate downhill towards the river.

The city was busy, in spite of the weather. Black and red, and blue and yellow umbrellas formed a canopy over the heads of shoppers as they flowed like rainwater from Bath Street and Renfield Street into the town's most famous shopping boulevard. Sauchiehall Street, deriving its name from the old Scots word *sauchiehaugh*, which roughly translated as *willow grove*. A meadow once filled with trees. A far cry from the tall steel and brick buildings, and the few remaining red sandstone tenements that lined it now.

Addie passed the 3D cinema complex on the corner and glanced up for the first time towards the top of the hill, and the glass tower at the far side of the square formed by the buildings of the Scottish International Media Consortium. The home of Charles Younger's newspaper, the *Scottish Herald*. Although most of its publishing these days was conducted online, the *Herald* still produced a daily newspaper. Its circulation amounted to only a few thousand, but it was read by the country's top business people, its politicians and regulators, and by most in the legal profession.

She felt fear form a fist in her belly.

Robbie's old weekend pack, which her father had been wearing when he was shot, weighed heavily on her shoulders, chafing at them even through the layers of her parka. His blood stained the inside face of it, but was not visible to the casual eye.

She could feel the temperature falling, even as she crossed the street. It was forecast to dip below freezing, with wet roads and pavements turning to ice in the coming hours.

The previous evening she had spoken to the newspaper's editor, Richard Macallan, for less than ten minutes, from one of the few remaining public telephone booths on the south side. The incoming call to the *Herald*, Tiny told her, would be monitored. It wouldn't take them long to trace the source of it. But unless Macallan knew she was coming, she would never get past security. So the phone call was necessary to alert him. But the authorities would have been alerted, too. And she could only stay on the line for a few minutes before they would come looking for her.

Now, she knew, they would be waiting for her at the top of the hill.

She skirted the traffic barrier and walked up into the tiny square, which was more of a turning circle, built around a unicorn raised on a tall pillar above an old stone fountain. The unicorn: Scotland's national, mythical, animal. A symbol of purity and innocence. A sad irony, given Addie's reason for being there today.

Cars stood parked in a row along the left side, and two men in dark suits and long raincoats emerged from a black Merc. They strode quickly across the cobbles to intercept her. And were exactly as Addie had imagined, living out some comic book fantasy of their own importance.

Both men were startled by the sound of tyres skidding on wet cobblestones, and they turned to see four marked police vehicles speeding through the raised barrier. The cars divided to flank the tiny group in the circle, and flak-jacketed police officers poured out, Heckler & Koch MP5 sub-machine guns levelled at the men in raincoats.

Almost by instinct, the two men reached for concealed weapons beneath their coats, but stopped as a tall, plain-clothes officer emerging from the lead car barked at them, 'Remain perfectly still, or you will be shot where you stand.'

'Do you have any idea what you're doing?' the older of the two demanded, angry spittle gathering around his lips.

'Yes, sir, I do,' Tiny said. He felt the rain dripping from his nose and chin. 'We're responding to a tip-off about a terrorist attack on the offices of the *Scottish Herald*. Now, lay your weapons very carefully on the ground in front of you.' Both men responded, gingerly removing Glock 26 pistols from leather holsters, to place carefully on glistening wet cobbles.

One of the uniformed officers moved in to pick them up, then retreated. 'Now show me some ID.'

The one who had spoken reached into an inside pocket.

'Careful!' Tiny warned him, and the man moved more cautiously to produce a leather wallet, which he flipped open and held out towards the policeman. Tiny approached to take it from him. He looked up, surprised. 'SIA?' And cast a doubtful look from one to the other. 'What are you doing here?'

The two men exchanged glances. And after a pause, 'Same as you,' said the older one.

'Oh, aye?' Tiny's eyes narrowed doubtfully. 'How come we weren't informed?'

The man shrugged. 'Crossed wires, I guess.'

Tiny handed him back his wallet. 'We're going to have to check you out. You'll come with us.' And neither of them was going to argue with him.

As they were led to the nearest vehicle, one of them glanced back towards the entrance to the *Herald*. The girl was nowhere to be seen.

Addie stepped out of the elevator and followed the young woman through a busy newsroom. A few heads lifted from computer screens to glance curiously in her direction. The girl opened a glass door into a fishbowl of an office with windows all along the far side, and ushered Addie in.

Macallan was a man of about Brodie's age. He had a sculpted face with wary dark eyes, and the remains of once fair and abundant hair gelled back across a broad skull. He stood up from his desk and held out a bony hand, which Addie shook tentatively. He said, 'I watched that whole debacle down there from the window. You must have friends in high places.'

Addie said, 'My father had friends who owed him a favour.'

'What have you got for me?'

Addie swung the pack from her shoulders to set on his desk. 'Everything.' She unzipped it to bring out Younger's laptop, his notebooks and printouts, and the report which had sparked off his whole investigation.

Macallan lifted the A4 ring binder and flipped through the pages of shorthand notes. He lifted a hand to wave someone through from the newsroom and pushed all the notebooks towards the young journalist who entered. 'I want all this stuff transcribed, as soon as possible. As many people on it as it takes.' He picked up the report then and shook his head in wonder as he riffled through it. He looked at Addie. 'You know if this all holds up, it'll bring down the government.' He sighed. 'Of course, they'll claim that any publication of Younger's story is in breach of standing DSMA-Notice regulations.'

Addie said, 'I don't know what that is.'

'The Defence and Security Media Advisory Committee decides what is in the public interest, and what is a danger to national security.'

'They murdered my father.' Addie stared unblinking at the editor. 'And your journalist.' She delved into the pack again and pulled out Sita's notebook. 'The pathologist's notes on his autopsy, before they murdered her, too.'

Macallan looked at his watch. 'We've got ten hours to make this stand up. And if we do, I'll publish. Then I'll fight them in the courts if I have to.' A pale smile flitted across his face. 'Better to be forgiven than forbidden.'

CHAPTER THIRTY-THREE

The SEC Armadillo was jam-packed. A sea of waving flags and banners. The chanting of the crowd rising to the rafters, reverberating throughout the auditorium.

The stage was bedecked by elongated saltires hanging from the roof, the campaign logo of the Scottish Democratic Party projected in blue on the screen behind the podium. ONWARD TO VICTORY.

Sally Mack was an island of calm in the eye of the storm. She stood at the podium smiling, facing a barrage of media mics. She turned her head slowly from one invisible teleprompter to the other, delivering carefully considered words crafted by half a dozen speechwriters. Victory was theirs. The future of Scotland assured. Tomorrow the electorate would return to power the party that had delivered both economically *and* ecologically.

She was a slim and elegant woman in her early sixties, her calf-length blue dress emphasising both her femininity and her power. Her carefully sculpted and dyed blond hair made Addie think of the photographs she had seen of the first woman to become British prime minister, Margaret Thatcher. Her delivery had the same syrupy sense of insincerity. Here, she told her adoring crowd, stood the woman who had delivered energy certainty for Scotland, while most of the rest of the world was still struggling to come to terms with the post-fossil fuel emergency. And suffering the consequences of their failures.

Addie and Sheila sat together on the edge of the settee, watching the screen, taut with tension. Sally Mack's triumphalism was both infuriating and depressing. Addie wanted to throw something at her. Anything. But she contained her frustration. She was exhausted after the hours of intensive grilling she had endured at the offices of the *Herald*.

Cameron, wrapped in a blanket, was asleep in Tiny's armchair. Oblivious. Addie glanced away from the screen towards her son, and her heart and soul bled for him. Just a matter of days ago, they had been the picture-perfect family, living the dream in one of the most beautiful parts of Scotland, perhaps the world. And it had all been an illusion. The dream, a nightmare just waiting for the hours of darkness. And the darkness, when it came, had been both bloody and profound.

Addie had barely slept since the moment of pulling the trigger and watching the man she had once loved thrown backwards into the snow. The same look in his eyes then as she had seen in Brodie's just thirty minutes later.

But she had no more tears to cry. They had spilled until she ached, her eyes red and scratchy, burning now only with anger.

She almost jumped at the sound of the front door opening, and Sheila leapt immediately to her feet. Tiny appeared in the living room door, his face grey and drawn. His overcoat hung limp from bony shoulders, dripping rainwater on the carpet.

'Are you okay?' Sheila's voice was tentative.

Tiny sighed. 'It's been a long day. And I'd probably have been in a lot more trouble if those SIA guys had been able to claim they were there on official business.' He slipped off his coat to hang on the coat stand in the hall, and they heard his voice come back to them from over his shoulder. 'As it was, they had to go along with our story of a terror warning to explain why they *were* there.' He came back into the room. 'But it's a mess. And I'm not out of the woods yet.' He managed a pale smile. 'Though I think we're going to be okay.'

'What's SIA?' Addie said.

'Scottish Intelligence Agency, pet. Not that there was much intelligence discernible in those guys.' He disappeared into the kitchen to open the fridge and grab a beer. As he came back through to the living room, he popped the lid off the bottle and raised it to his lips. He took a long draught. Then he said, 'The good news is that the *Herald* have published. Simultaneously on the internet and in print.' He mimicked the sensational delivery of an imagined newsreader. '*Herald* reporter murdered to cover up disastrous radiation leak at Ballachulish A.'

He slumped into the vacant armchair.

'Everyone's picking up on it. It'll be the lead story on every news bulletin all day tomorrow, and probably for weeks to come. Trust me, there's not an elector in the land who won't have seen it before they go in to cast their vote.'

Their attention was suddenly drawn to the TV as an announcer's voice broke across coverage of the SDP rally at the Armadillo. 'We have breaking news.' Simultaneously, a BREAKING NEWS banner appeared, and an inset of a station newsreader popped up in the bottom left corner of the screen with news of the sensational story just published by the *Scottish Herald*.

The director covering the rally cut to a close-up of the podium as a po-faced man in a dark blue suit whispered into the ear of the first minister, whose strained smile could barely conceal her irritation at this on-stage interruption at the climax of her speech.

But the smile very quickly vanished, and Sally Mack's mouth gaped just a little, initially shocked. Before fear and realisation registered in the widening of her eyes. Game over.

Addie punched the air in vengeful satisfaction. 'Yes!'

CHAPTER THIRTY-FOUR

Addie walked with Cameron through the old Cathcart Cemetery, fallen leaves crackling underfoot in the frost. Here stood the graves of the good and the great. Impressive headstones and mausoleums. Wonderful old trees bowed in reverence by time and death, witness to the passing of generations.

It was deserted on this icy December day, a pale disc of winter sun barely rising above the southern hills of the city.

The little boy clutched his mother's hand, swaddled in clothes to keep him warm, red nose in a bright face beneath his yellow woollen bunnet, talking almost incessantly about his day out in Pollock Park just yesterday with Uncle Tony and Auntie Sheila. They had taken him horse riding, and then to a café for ice cream, in spite of the cold. He wasn't complaining. And they, Addie reflected, seemed almost reborn. Happy to take on responsibility for a family they'd never had. Not, she knew, just out of Tiny's loyalty to her dad, but because they wanted to.

Addie led Cameron down the path to Netherlee Road and they crossed to where the cemetery had been extended into the Linn Park. It was more open here. Less mature. And they found her dad's grave easily among the rows of recent headstones. Placed in the ground close to where he had buried his wife ten years earlier.

There was a wooden bench on the edge of the path, and after she had laid her flowers on the grave, she lifted Cameron on to it and sat down beside him, staring at the simple inscription on the headstone.

Cameron Iain Brodie, 5th April 1996 to 23rd November 2051. Loving husband and father.

The time for tears was long gone, but the regret would linger a lifetime.

Cameron said, 'I don't know why my grampa had to die? Just when we found him.' He thought about it. 'Everyone else has a grampa. Some of the boys at school even have *two*.' His sense of wonder at this was expressed in the emphasis he placed on the word. 'You know what I wish, Mum?'

'No, Cammie, what do you wish?'

'I wish Grampa didn't *have* to go to heaven.'

Addie pressed her lips together to contain her emotion. 'Me too, Cammie. Me too.'

THE END

ACKNOWLEDGEMENTS

I would like to offer my grateful thanks to those who gave so generously of their time and expertise during my researches for *A Winter Grave*. In particular, I would like to express my gratitude to Dr Steve Campman, medical examiner, San Diego, California, USA, for his advice on forensics and pathology; Mo Thomson, photographer, whose amazing still and drone photography substituted for my eyes and ears in Kinlochleven and on Binnein Mòr, when Covid-19 made it difficult for me to travel. Mo's virtual eVTOL flights from Glasgow to Mull and through Glencoe to Loch Leven, as well as his simulated flights to the summit of Binnein Mòr and into the corries, provided stunning insights into the landscape; Professor Jim Skea, co-chair of Working Group III of the United Nations Intergovernmental Panel on Climate Change (IPCC), and member of the UK government Committee on Climate Change for his advice on my climate change scenario; Cameron McNeish, author, Scottish wilderness hiker, backpacker and mountaineer for his insights on snow, and climbing Binnein Mòr.